Lin Qiu, Yanhui Feng
Thermal Engineering

De Gruyter Textbook

—

Lin Qiu, Yanhui Feng

Thermal
Engineering

—

Engineering Thermodynamics and Heat Transfer

DE GRUYTER    Science Press
Beijing

Authors
Lin Qiu
University of Science and Technology Beijing
30 Xueyuan Road, Haidian District
Beijing 100083, China
qiulin@ustb.edu.cn

Yanhui Feng
University of Science and Technology Beijing
30 Xueyuan Road, Haidian District
Beijing 100083, China
yhfeng@me.ustb.edu.cn

ISBN 978-3-11-132969-7
e-ISBN (E-BOOK) 978-3-11-132970-3
e-ISBN (EPUB) 978-3-11-132975-8

Library of Congress Control Number: 2023950573

Bibliographic information published by the Deutsche Nationalbibliothek
The Deutsche Nationalbibliothek lists this publication in the Deutsche Nationalbibliografie;
detailed bibliographic data are available on the Internet at http://dnb.dnb.de.

© 2024 China Science Publishing & Media Ltd. and Walter de Gruyter GmbH, Berlin/Boston
Cover image: Kannikaistock2499/iStock/Getty Images Plus
Typesetting: Integra Software Services Pvt. Ltd.
Printing and binding: CPI books GmbH, Leck

www.degruyter.com

Preface

To meet the needs of talent training in the new era, this textbook combines the classical content of engineering thermodynamics and heat transfer with modern research advances, and expands the scope of knowledge about thermal engineering. We have selected appropriate exercises, which can cultivate students' ability to think independently and solve problems and meet the needs of professional talent training in depth and breadth.

The book consists of 13 chapters, which is categorized into two parts. The first part is composed of eight chapters, that is, Chapters 1–8. Chapter 1, as an introduction, introduces the classification and definition of different energy forms and outlines the research content of this book. Chapter 2 elaborates some basic concepts of engineering thermodynamics, such as thermodynamic system, thermodynamic state, state parameters, and thermodynamic cycle. Chapter 3 introduces the physical essence and mathematical expression for the first law of thermodynamics for open system and closed system and their applications. Chapter 4 presents the properties and the thermal processes of ideal gas. Chapter 5 focuses on the second law of thermodynamics, the Carnot cycle, and the entropy. Chapter 6 elaborates the thermal process and properties of water vapor and wet air. Chapter 7 introduces the power cycle of steam and gas and the piston internal combustion engine cycle. Chapter 8 introduces the refrigeration equipment and the efficiency calculation method. The second part focuses on the heat transfer, which consists of five chapters, that is, Chapters 9–13. Chapter 9 introduces the basic ways of heat transfer and important definitions. Chapter 10 elaborates on the theoretical basis of heat conduction and the relevant knowledge of steady-state heat conduction. Chapter 11 introduces the basic theory of thermal convection, that is, Newton's cooling law, and the influencing factors on heat transfer. Chapter 12 introduces the basic concept of thermal radiation, the basic law of blackbody radiation, the emission characteristics of actual objects, atmospheric greenhouse effect, and greenhouse effect. Chapter 13 describes the heat transfer process and heat exchangers, as well as the enhanced and weakened heat transfer approaches.

This book is of particular use to thermodynamic engineers, mechanical engineers, electrical engineers, and low carbon practitioners worldwide, as well as to academics and researchers in the fields of thermal management, energy engineering, and material science.

This book was financially supported by the textbook construction fund of the University of Science and Technology Beijing and the Academic Affairs Office of University of Science and Technology Beijing.

https://doi.org/10.1515/9783111329703-202

Acknowledgments

As the editor of this book, I am very grateful for the financial support from the planning textbook construction funds of the University of Science and Technology Beijing (no. JC2022YB013). The authors of each chapter also provided great help. I would like to express my sincere gratitude to the authors Yiling Liu, Xin Wang, Yijie Yang, Zhaoyi Wang, Yuxin Ouyang, Guangpeng Feng, and Zihan Liu. In addition, I also express my gratitude to the books cited. Most of the contents of this book are collected from various related journals and works. Finally, I would like to thank all the readers of this book. It is your support and affirmation that gave the motivation to publish. Owing to the relatively short writing time, omissions and even errors in the book are inevitable. We will listen carefully to readers' criticisms and correction.

https://doi.org/10.1515/9783111329703-203

Contents

Part 2: **Heat transfer**

Part 1: **Engineering thermodynamics**

Chapter 1
Introduction

Energy is the engine of social development because humans need various forms of energy in their daily lives. With the development of society, the cognition and utilization of energy are constantly improving. With the development of science and technology, the efficiency of energy utilization has been greatly improved, and our society has been developed into the Industry 4.0 era.

1.1 Energy

1.1.1 Definition of energy

The world is made up of matter. All matter is in motion. Energy is a measure of matter in motion. All matter has energy, which means that the world would be in stationary state without energy, and lives would be nowhere to be found. Corresponding to various forms of motion of matter, there are also various forms of energy, which can be converted to each other, but the total amount remains the same. Heat energy can be converted into electricity, which can be further converted into light energy. However, the total amount of energy is a constant.

1.1.2 Main forms of energy

Energy exists in numerous forms such as thermal, mechanical, kinetic, potential, electrical, magnetic, and nuclear, and their sum constitutes the total energy (E) of a system. There are six main forms of energy that are recognized and utilized nowadays:

1. ***Thermal energy***: This is the sum of the kinetic energy of the thermal motion of molecules and the potential energy due to the intermolecular interaction. Temperature reflects the intensity of thermal motion of molecules, which means that it is also the macroscopic reflection of substances' thermal energy. Thermal energy is widely used by human being at early times, such as previous drilling wood for cooking fire and the internal combustion engine now. As shown in Fig. 1.1(a), ice will melt and become water when it absorbs thermal energy from human hands.

2. ***Electric energy***: This is the energy related to the movement and the accumulation of electric charge. It is one of the extensively used energy forms. As shown in Fig. 1.1(b), electric energy can turn into lighting.

3. ***Mechanical energy***: Mechanical energy is the first recognized and utilized energy, which consists of kinetic energy and potential energy of an object such as

https://doi.org/10.1515/9783111329703-001

the water wheel during ancient times and the wind power engine at present. As shown in Fig. 1.1(c), the ball's kinetic and potential energy can convert to each other.

4. **Radiant energy**: This is the energy emitted by an object in the form of electromagnetic waves. Radiant energy is usually converted from other forms of energy. As shown in Fig. 1.1(d), the Sun's radiation energy can burn through an aluminum plate.

5. **Nuclear energy**: Nuclear energy refers to the energy generated by nuclear reaction. It is of great importance in the future for electricity production. Fig. 1.1(e) is a picture of a model of nuclear power plant.

6. **Chemical energy**: Chemical energy refers to the energy released through chemical reactions. Various fuels, including coal, gas, and oil, are common examples of the sources of chemical energy. As shown in Fig. 1.1(f), wood contains chemical energy, which will be released by combustion.

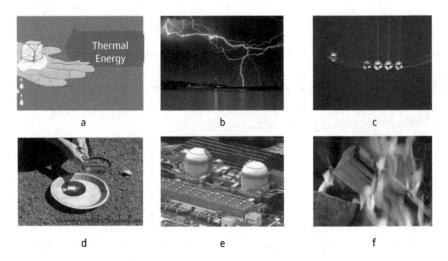

Fig. 1.1: Six representative energy forms: **(a)** thermal energy, **(b)** electric energy, **(c)** mechanical energy, **(d)** radiant energy, **(e)** nuclear energy, and **(f)** chemical energy.

1.1.3 Properties of energy

1.1.3.1 Quantitative attribute: first law of thermodynamics

All kinds of energy are equal and have additive property, with no distinction of its quantity.

The first law of thermodynamics shows that energy can either be created or destroyed, suggesting the conservation of energy during its transfer and transformation process, which is one of the fundamentals of thermodynamic analysis.

1.1.3.2 Qualitative attribute: second law of thermodynamics

The quality of energy can only go from high to low, and it is impossible to go from low to high spontaneously.

The second law of thermodynamics reveals the direction, conditions, and limits of all thermal processes in nature.

1.2 Energy source

1.2.1 Definition of energy sources

Energy source refers to material resources that can provide energy directly or indirectly. Energy source is what human society relies on for survival and development. There are various forms of energy resources on the Earth, which are classified according to the original source, development steps, utilization, and whether they can be recycled.

1.2.2 Classification of energy sources

There are many classification criteria for energy source, generally including the following ways:

According to the initial source, energy source can be produced by extraterrestrial bodies, the Earth's interaction with other celestial bodies and the Earth itself. Energy produced from extraterrestrial bodies is mainly solar energy. Fossil fuels (coal, oil, and natural gas) are essentially solar energy fixed by ancient organisms. In addition, water energy, wind energy, and wave energy are also converted from solar energy. Energy from the Earth's interaction with other celestial bodies includes tidal energy. Energy from the Earth itself includes nuclear energy and geothermal energy.

According to the socioeconomic status, it can be classified into conventional energy and new energy. Conventional energy is the energy with long exploitation time, mature technology, and wide usage. New energy is the energy that has not been developed and utilized on a large scale due to immature technology and short exploitation time.

According to the development steps, it can be classified into primary energy source and secondary energy source. Primary energy source exists in pristine form in nature, which can be directly exploited and utilized. Secondary energy source is the energy directly or indirectly converted from primary energy.

According to whether the energy source can be regenerated or not, it can be classified into renewable energy source and non-renewable energy source. Renewable energy source would not be significantly reduced by exploitation and have a natural resilience. Non-renewable energy source has limited reserves, and its amount decreases with the development and utilization. Eventually, it will be exhausted.

According to the environmental pollution in the process of development and utilization, it can be classified into clean energy source and non-clean energy source. Clean energy source has no or little pollution to the environment. Non-clean energy source would cause great pollution to the environment.

The most common energy forms in nature are shown in Fig. 1.2; most of which belong to thermal energy and mechanical energy category. Thermal energy occupies the dominant status. Mechanical energy mainly includes hydraulic energy, wind energy, and tidal energy. According to the type of energy release reaction, thermal energy can be released by combustion and fission (or fusion), where chemical energy can be released by combustion, such as coal, oil, natural gas, and biomass, while nuclear energy, solar energy, and geothermal energy can be released by fission and fusion.

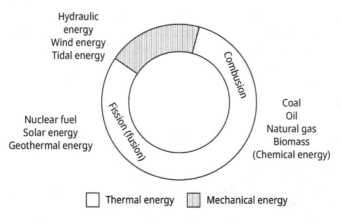

Fig. 1.2: Energy forms and their share in the nature.

With the development of human society, there are five major issues of global concern: energy, population, food, environment, and resource. Simultaneously, 3E problem is also proposed. 3E problem refers to energy, environment, and economy. Therefore, energy construction is one of the strategic emphases worldwide.

According to **BP Statistical Yearbook of World Energy** 2021 [1], COVID-19 has exerted a huge impact on energy markets, with primary energy and carbon emissions decreasing at the highest pace since World War II. Among this, the global primary energy consumption has dropped by 4.5% in 2020, with oil accounting for three-quarters roughly, which is the biggest drop since 1945. In spite of this, the demand for renewable energy like wind energy, solar energy, and hydroelectric power continues growing, with solar energy making history. On a country-by-country basis, the United States, India, and Russia saw the biggest drops in energy consumption. China, with the largest increase (2.1%), was one of the few countries whose energy demand increased in last year. We will elaborate the status of world energy as follows:

Carbon emission: Carbon emission due to energy use has dropped by 6.3%, which is the lowest since 2011. In terms of primary energy, it was also the biggest drop since the end of World War II.

Fossil oil: The largest regional oil demand reduction occurred in the United States (decrease by 2.3 million barrels/day (b/d)), followed by the European Union (decrease by 1.5 million b/d) and India (decrease by 480,000 b/d). In fact, China is the only country that renders an increase in oil consumption (220,000 b/d). Global oil production has decreased by 6.6 million b/d, with OPEC accounting for two-thirds of the decline. Libya (decrease by 920,000 b/d) and Saudi Arabia (decrease by 790,000 b/d) had the largest declines among OPEC members. Moreover, among non-OPEC members, Russia (decrease by 1 million b/d) and the United States (down 0.6 million b/d) saw the largest declines.

Natural gas: Natural gas consumption has decreased by 81 billion cubic meters (BCM), or 2.3%. In contrast, the proportion of natural gas in primary energy continued to rise, reaching a record of 24.7%. Russia (decrease by 33 BCM) and the United States (down 17 BCM) are the countries with the biggest decline in gas demand in 2020, while China (22 BCM) and Iran (10 BCM) saw the biggest increase.

Coal: Coal consumption has decreased by 6.2 EJ, or 4.2%, with the United States (decrease by 2.1 EJ) and India (decrease by 1.1 EJ) decreasing the most. China and Malaysia were two notable exceptions, with coal consumption increasing by 0.5 EJ and 0.2 EJ, respectively.

Renewable energy, hydropower, and nuclear power: Renewable energy including biofuels but excluding hydropower has increased by 9.7%, slower than the average of the past decade (13.4% per year), but the increase in energy (2.9 EJ) was similar to that in 2017, 2018, and 2019. Solar power has increased by 1.3 EJ (20%), making a new record, with wind power accounting for the largest increase in renewable energy (1.5 EJ). The installed solar capacity has increased by 127 GW. Wind capacity has increased by 111 GW, which is almost twice of the largest increase in previous years. China had the largest increase in renewable energy (1.0 EJ), followed by the United States (0.4 EJ). European Union has increased by 0.7 EJ. Hydropower has increased by 1.0%, of which China has increased the most (0.4 EJ). Nuclear power has decreased by 4.1%, with the biggest decline in France (−0.4 EJ), the United States (−0.2 EJ), and Japan (−0.2 EJ).

Electric power: Global electricity production has dropped by 0.9%, surpassing the 0.5% decline in 2009, which is the only decrease in records since 1985. The proportion of renewable energy generation increased from 10.3% to 11.7%. Meanwhile, the proportion of coal generation has dropped by 1.3–35.1%, which is BP's lowest record.

Important minerals: Global lithium production has dropped by 4.6% due to capacity reduction in Australia. Cobalt production in the Democratic Republic of Congo declines in 2019, but increases by 2.9% in 2020 as the country's capacity partially restores. Influenced by substantial growth in Australia and the United States, the production of rare earth metals has increased by 23.2%.

Figure 1.3 illustrates the proportion of primary energy consumption for different regions in the world in 2020, where China's share is 27.2%.

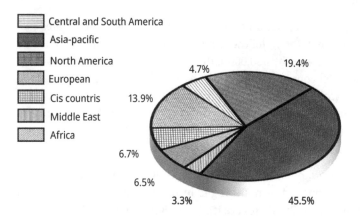

Central and South America
Asia-pacific
North America
European
Cis countris
Middle East
Africa

4.7%   19.4%

13.9%

6.7%

6.5%

3.3%   45.5%

Fig. 1.3: Percentage of global primary energy consumption for different regions in year 2020.

As shown in Fig. 1.4, in 2020, energy producers overcome the negative impact of COVID-19 and actively promote the resumption of work and production, with a total energy production of 4.08 billion tce (tons of standard coal equivalent) and a year-on-year increase of 2.8%. Among them, raw coal output is 3.9 billion tons, with an increase of 1.4% year on year. Crude oil production is 19.4769 million tons, with an increase of 1.6% year on year. Natural gas production is 192.5 billion m^3, with an increase of 9.8% year on year. Power generation is 7,779.06 billion kW · h, with an increase of 3.7% year on year.

As shown in Fig. 1.5, with the recovery of China's economy and social order, energy consumption has also showed a gradual recovery trend. In 2020, the total energy consumption was 4.98 billion tce, with an increase of 2.2% over the previous year, among which coal has increased by 0.6%, crude oil by 3.3%, natural gas by 7.2%, and electricity by 3.1%.

Besides, energy production and the consumption structure continued to get improved, as shown in Fig. 1.6. Driven by a series of policies and measures such as deepening energy supply-side structural reform and giving priority to the development of non-fossil energy, China's clean energy continues to develop rapidly. The proportion of clean energy has further increased, and the energy structure has been optimized continuously. In the past decade, the proportion of different types of energy has different trends. The share of raw coal and crude oil production continues to decline, while the share of natural gas production does not change much. The proportion of clean electricity production, including hydropower, nuclear power, wind power, and solar power, has increased significantly. In 2020, clean electricity production accounts for 28.8% of the total electricity generation.

The energy consumption structure in China is shown in Fig. 1.7. Similarly, the proportion of coal consumption shows a declining trend, accounting for 56.8% of the total

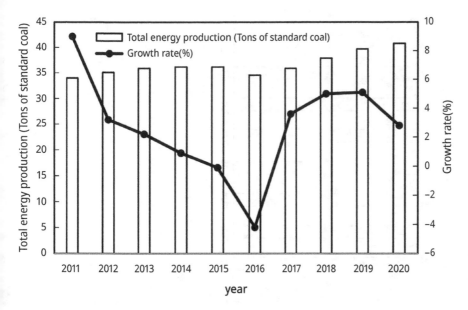

Fig. 1.4: Total energy production and its growth rate.

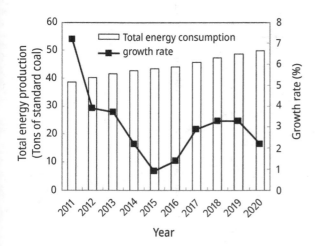

Fig. 1.5: Total energy consumption and its growth rate.

energy consumption in 2020, but it is still the main energy source in China in short term. The share of clean energy consumption, including natural gas, hydropower, nuclear power, and wind power, in total energy consumption has increased from 13% in 2011 to 24.3% in 2020. In general, coal plays a dominant role in China's energy structure, oil and natural gas are highly dependent on the import from foreign countries, and the proportion of clean energy consumption increases continuously.

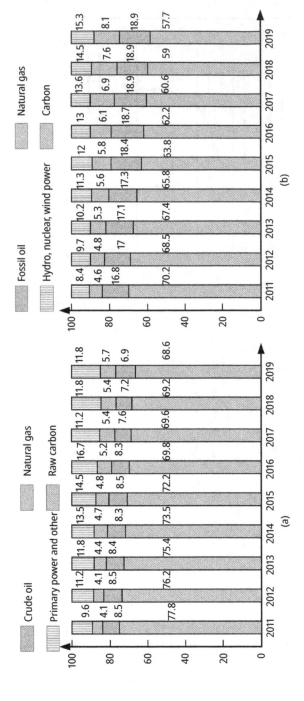

Fig. 1.6: **(a)** Energy production structure and **(b)** energy consumption structure in 2011–2019.

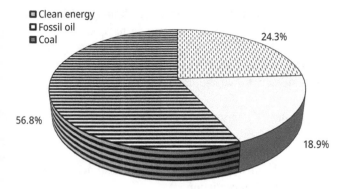

Fig. 1.7: Energy consumption structure in 2020.

Moreover, the energy efficiency level continues to improve. In recent years, China has made efforts to save energy and reduce carbon emissions to realize clean and efficient use of energy. Energy consumption and carbon emissions continue to decline. Energy imports have maintained rapid growth, especially for oil and gas resources. Renewable energy has been developed rapidly. The total installed capacity of renewable energy power generation in China has reached 930 million kW · h, with an increase of about 17.5% year by year and 15.7% compared to year 2011, accounting for 42.4% of the total installed capacity. Renewable energy generation has reached 220 million kW · h, with an increase of about 8.4% year by year, which quadruples the current growth rate of total energy consumption.

There are many challenges and difficulties in China's energy development. The first is the low per capita energy resources, which is 62% of the world average level in coal and 7% in fossil oil and natural gas. The second is that energy security is difficult to ensure. From 1.6% in 1993 to 72% in 2019, the oil's dependence on foreign countries has been increasing, and the crude oil import source is still intensive. Similarly, natural gas import has increased rapidly, and the dependence on foreign countries has reached 42.8% in 2019. Besides, the priority to set prices for oil, natural gas, and other products is still in the control of developed countries. Although the energy import channel has been gradually improved, there are still some risks. There is still a long way to go forward for the development of new energy.

In order to keep sustainable development of energy construction in China, *Outline of the national medium- and long-term plan for science and technology development* was issued, which claims that we should adhere to the priority of energy saving and energy consumption reduction, promote the diversification of energy structure, promote clean and efficient utilization of coal and reduce environmental pollution, strengthen the digestion, absorption, and re-innovation of imported technology of energy equipment, and improve the technical ability of regional optimal allocation of energy. In a word, for sustainable development of energy construction, we can find methods from reducing expenditure, broadening sources, saving energy, and reducing emissions.

New energy refers to new kind of energy and new energy technology. New energy technology is the study hot spot for engineers. Five elements of new energy technology are:
- High efficiency: significant improvement in thermal efficiency
- Environment friendly: zero emission, no vibration, low noise technology
- Industrial scale
- Life cycle analysis of the feasibility of energy
- Economic feasibility

Some new energy technologies have been widely used in practice and achieved good results in energy conservation and emission reduction.

1.2.3 Hydrogen energy and fuel cell technology

According to the study in 2021 [2], a fuel cell is an energy conversion device essentially consisting of an anode, electrolyte, and a cathode, which can directly convert chemical energy into electrical energy with higher efficiency and lower emissions.

Hydrogen is quite abundant in nature. There is no carbon dioxide or air pollutants' by-product emission but water when hydrogen is electrochemically oxidized in a fuel cell system, making hydrogen a clean energy. Hence, hydrogen can be used in many fields, such as transportation, power generation, and militarized equipment [3].

According to the type of electrolyte, fuel cells can be classified as alkaline fuel cells (AFCs), proton-exchange membrane fuel cells (PEMFCs), phosphoric AFCs, molten carbonate fuel cells, and solid oxide fuel cells. PEMFC stands out from all types of fuel cell due to its unique advantages of rapid start-up time, wild range of operating temperature (−40 to 90 °C), and high specific energy. PEMFC has PEM directly affecting its performance, which has functions of conducting protons, separating fuel oxidizer, and insulating protons. It provides a channel for the migration and transport of protons so that protons can pass through the membrane from the anode to the cathode and form a circuit with the electron transfer of the external circuit to provide the external current. It plays a very important role in the performance of the fuel cell, and its quality directly affects the service life of the battery [4].

Figure 1.8 depicts the working principle of a fuel cell, which occurs when oxygen and hydrogen are used to produce water, electrical power, and heat [3, 4]. There are various designs available for fuel cells, whose difference mainly origins from the difference in the chemical characteristics of the electrolyte [5]. It is to be noted that they all operate with the same basic principles and generate electricity and heat through electrochemical reaction.

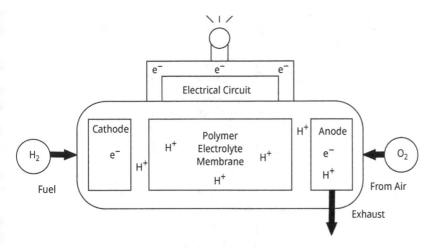

Fig. 1.8: Schematic diagram of the working principle of a fuel cell.

1.2.4 Advanced coal to chemicals industry (ACCI)

According to a study in 2021 [6], the advanced coal to chemicals industry (ACCI) is an important approach to use coal in a clean, highly efficient, and low carbon manner, with the purpose to solve China's coal-dominated energy structures, excessive imported oil and natural gas, and strict environmental constraints. The 12th Five-Year Plan saw constantly improved technologies and equipment. However, there are several challenges as well, such as insufficient strategic understanding, severe external constraints, and immature technology.

The industrialized technology routes of ACCI include preparing gas and liquid fuels and chemicals. Generally, it uses coal as the feedstock to synthesize natural gas and conventional oil-based products, such as gasoline, diesel, olefins, aromatics, and ethylene glycol, which is conducted via pyrolysis, gasification [7], liquefaction, and downstream processes, and thus partly replaces oil and natural gas. Compared with the processes used in the traditional coal chemical industry, ACCI utilizes advanced conversion technology and thus has long and complex processes. Figure 1.9 illustrates that the gasification approach starts from coal gasification technology, and then uses syngas as an intermediate product to produce methanol, Fischer–Tropsch oil (e.g., naphtha, gasoline, diesel oil, liquefied petroleum gas, etc.), synthetic natural gas, and ethylene glycol. Methanol can be further converted into olefins, gasoline, and aromatics. The liquefaction approach refers to the direct coal liquefaction process, whose main products are diesel and jet fuel. The pyrolysis approach is the deep processing of coal tar produced via medium- and low-temperature pyrolysis.

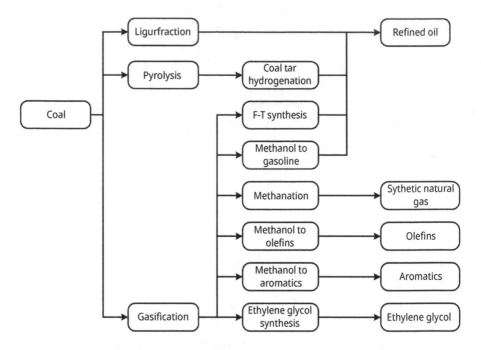

Fig. 1.9: Technology routes for ACCI in China.

1.2.5 Solar photovoltaic technology

According to a study in 2017 [7], photovoltaic (PV) cells, which is also known as solar cells, are made from the same semiconductor materials used in electronics and computer chips. The diversity of PV materials, different potential characteristics, and low-cost, versatile fabrication technologies are the merits of this energy conversion system. Solar cells are essentially composed of large-area PN junctions, which are made of the combination of a p-type semiconductor and an n-type semiconductor to form a junction. Asymmetric doping of the p-type and n-type semiconductor creates a potential barrier to the major charge flow between the regions. This results in a depletion region where the flow of electrons and holes ceases in equilibrium state, and the chemical forces of diffusion are exactly balanced by the electrostatic forces of the created electric field. The concentration gradient provides a necessary electronic asymmetrical condition for PV cells.

Figure 1.10 illustrates the operational principle of a solar cell [7]. When light is illuminated onto a piece of solar cell, photons of different wavelengths hit the semiconductor surface. Only a fraction of photons are converted into electrical energy as only photons with energy higher than energy band gap of the semiconductor can be absorbed. Photon absorption leads to generation of electron–hole pair (EHP). The majority carrier concentrations (the total number of electrons in an n-type semiconduc-

tor or the total number of holes in a p-type semiconductor) are not affected by contributions from the additional photons because concentrations that EHPs generate are insig-
nificant compared to the majority carrier concentrations. However, minority carrier
concentrations (the total number of electrons in a p-type semiconductor or the total
number of holes in an n-type semiconductor) are affected significantly and experience
an increase. This change upsets the equilibrium condition between the diffusion force
and electrostatic force. Electrons originated from the p region eventually diffuse into the
depletion region, where the potential energy barrier at the junction is lowered, allowing
current to flow and establish a voltage at the external terminals. Holes created in the
n-doped region travel in the opposite direction to the p-doped side. Solar cells are
based on the movement of charges in the above manner, which will finally produce
electrical current.

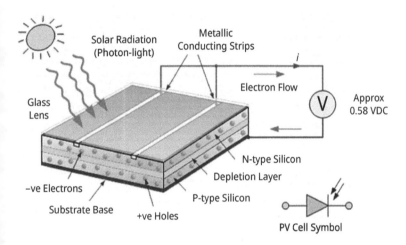

Fig. 1.10: Working principle of solar photovoltaic devices.

1.2.6 Distributed energy supply technology

The distributed energy source is a comprehensive energy utilization system which is
close to the client. In the transmission and utilization of energy, it is arranged in
pieces to reduce the loss of long-distance transmission of energy, and effectively improve the safety and flexibility of energy utilization.

Distributed energy supply technology refers to technology with the following
characteristics: (1) the primary energy is mainly composed of gas fuel, supplemented
by renewable energy; (2) the secondary energy mainly consists of heat, power, and
cooling cogeneration distributed at the user end, supplemented by other central energy supply systems, so as to realize the graded utilization of energy that directly
meets the various needs of users. Distributed energy systems utilize a variety of en-

ergy sources, including natural gas, biomass, wind, solar, and geothermal energy. In addition, it can be coupled with other energy forms such as waste heat, pressure, and gas. Because of the different energy forms, distributed energy systems differ in forms and structures. It mainly includes cogeneration of heat and power, renewable energy, energy storage, and fuel cell. The technical block diagram is shown in Fig. 1.11.

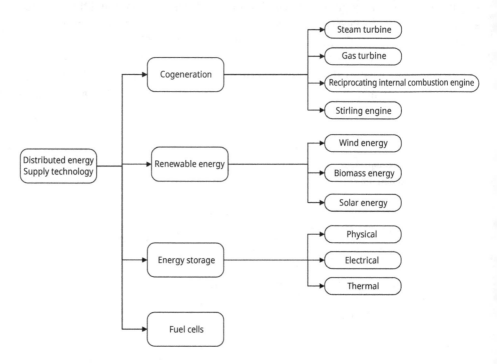

Fig. 1.11: Classification of distributed energy supply technologies.

1.3 Conversion and utilization of energy

As shown in Fig. 1.12, wind energy and hydraulic energy can turn into mechanical energy through mechanical devices, and mechanical energy can turn into electrical energy through generators. The chemical energy in hydrogen and alcohol can be converted directly into electricity by fuel cells. Geothermal energy can be used directly for heating. Solar energy can be converted into biomass energy and thermal energy through photosynthesis and solar collectors, respectively. In addition, other energies, such as chemical energy in coal, oil, and gas and nuclear energy, are usually converted into thermal energy directly or indirectly through combustion or nuclear reactions.

According to statistics, the energy utilized by thermal energy accounts for more than 90% of the total energy utilization in China and more than 85% for the other countries in the world. Therefore, in the process of energy conversion and utilization, ther-

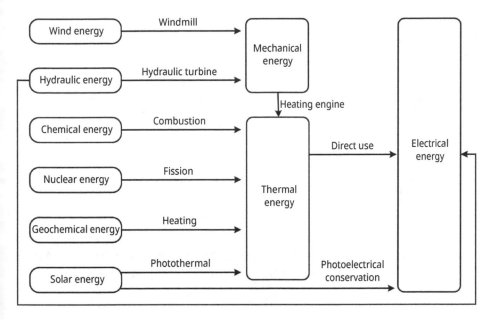

Fig. 1.12: Schematic diagram of the conversion of various energy forms.

mal energy is not only the most common form but also plays an important role. The effective utilization of thermal energy is of great significance to solve the energy problem of our country and even to the development of human society. There are two basic ways to use thermal energy, one is heat utilization, such as cooking, heating, drying, and smelting, and the other is power utilization. The thermal energy from heat engine can be changed to mechanical energy by generators and finally transformed into electric energy. Thermal efficiency of heat engine is improved with the development of technologies. For early steam engine, the thermal efficiency is 1–2%. Modern gas turbine plant has been lifted to 37–42%, and the efficiency of steam power station is approximately 40%. The use of thermal energy contains thermal power plant and refrigerating plant. Thermal power plant includes steam power plant, internal-combustion engine, gas turbine plant, and nuclear power plant. Refrigerating plant includes refrigerator and air conditioner.

1.3.1 Thermal efficiency

Thermal efficiency refers to the ratio of the effective energy output to the input energy for a specific thermal energy conversion device. Thermal efficiency is a dimen-

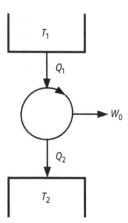

Fig. 1.13: Schematic diagram of a thermodynamic cycle.

sionless quantity, usually expressed as a percentage. The thermal efficiency of a thermodynamic cycle (Fig. 1.13) can be calculated by the following equation:

$$Q_1 = Q_2 + W_0 \tag{1.1}$$

$$\eta_t = \frac{W_0}{Q_1} \tag{1.2}$$

In an engine, the ratio of the amount of heat produced by mechanical work to the amount of heat absorbed by the boiler is calculated. The thermal efficiency of engine is divided into two kinds: indicative thermal efficiency and effective thermal efficiency. Indicative thermal efficiency refers to the ratio of the actual cycle work of the engine to the fuel equivalent heat consumed. Effective thermal efficiency refers to the ratio of the effective work in the actual cycle to the heat consumed, which is an important index to measure the economic performance of an engine.

1.4 Research contents of fundamentals of thermal engineering

Fundamentals of thermal engineering mainly consist of two parts: *engineering thermodynamics* and *heat transfer*. It mainly introduces the basic law of the utilization of thermal energy, the process of utilizing heat energy, and the basic law of heat transfer.

1.4.1 Engineering thermodynamics

Engineering thermodynamics is a branch of thermodynamics and a specific application of thermodynamics theory in engineering. Its contents include the law of mutual conversion between heat energy, mechanical energy and other forms of energy, and ways and technical measures to improve the efficiency of energy conversion.

The first law of thermodynamics and the second law of thermodynamics are the theoretical basis of engineering thermodynamics. The mutual conversion between heat energy and mechanical energy is realized through the state changing of working medium in thermodynamic cycles.

The macroresearch method of classical thermodynamics is adopted, and the abstract, generalization, idealization, and simplification methods are also adopted. Based on the first law of thermodynamics and the second law of thermodynamics, it studies the macroscopic thermodynamic process without considering the microscopic structure and circumstantial behavior of matter, so the results of the analysis and inference are reliable and universal.

1.4.2 Heat transfer

Heat refers to the energy transferred under the action of temperature difference. If there is a temperature difference, there will be heat transfer. Heat transfer is indispensable for all heat utilization processes. Temperature difference exists everywhere in nature, people's daily lives, and production practice, so heat transfer is a universal physical phenomenon, with a wide and profound impact on people's life and production practice. Therefore, it is of great practical significance to understand the law of heat transfer and master the methods and technical measures to control and optimize heat transfer. The research methods of heat transfer mainly include theoretical analysis, numerical simulation, and experimental research.

The learning purpose of engineering thermodynamics includes mastering basic concepts, laws and analysis methods, the law of the interconversion between thermal and mechanical energy, methods and technical measures to improve conversion economy, and then establish the concept of energy conservation and rational use.

The learning purpose of heat transfer includes mastering the basic concepts, theories, analysis methods, calculation, and experimental methods, and lays the necessary technical theoretical foundation for future research, treatment, and solution of practical heat transfer engineering problems.

Exercises

1. Categorize the following energy sources and fill in the letters on the lines.
   i. Conventional energy: _____
   ii. New energy: _____
   iii. Renewable energy: _____

    iv. Non-renewable energy: _____

    v.   Clear energy: _____

    vi. Non-clean energy: _____

    A. Wind   B. Solar energy   C. Hydro   D. Biomass

    E. Coal   F. Nuclear energy   G. Petroleum

2. What are the five major issues of global concern? What is the 3E problem?

3. What are the important elements of new energy technology?

4. What is the main research content of *fundamentals of thermal engineering*?

5. Engineering thermodynamics is not based on the first law of thermodynamics and the second law of thermodynamics. (True or False?)

6. Engineering thermodynamics adopted the macroresearch method of classical thermodynamics. (True or False?)

7. Heat transfer studies the macroscopic thermodynamic process without considering the microscopic structure and circumstantial behavior of matter. (True or False?)

8. Heat transfer studies the law of heat transfer. (True or False?)

9. Indicative thermal efficiency is an important index to measure the economic performance of an engine.

Answers

1. i. EG   ii. ABCDF   iii. ABCD   iv. EFG   v. ABCDF   vi. EG

2. Energy, population, food, environment, and resource. 3E problem is energy, ecology, and economy.

3. High efficiency, environment friendly, industrial scale, life cycle analysis, feasibility of energy, and economic feasibility.

4. *Fundamentals of thermal engineering* mainly consist of two parts: *engineering thermodynamics* and *heat transfer* It mainly introduces the basic law of the utilization of thermal energy, the process of utilizing heat energy, and the basic law of heat transfer.

5. F

6. T

7. F (Engineering thermodynamics studies the macroscopic thermodynamic process without considering the microscopic structure and circumstantial behavior of matter.)

8. T

9. F (Effective thermal efficiency refers to the ratio of the effective work in the actual cycle to the heat consumed, which is an important index to measure the economic performance of an engine.)

Chapter 2
Basic concepts

2.1 Review: introduction

Definition of energy: The world is made up of matter. All matter is in motion. Energy is a measure of matter in motion and all matter has energy. Corresponding to various forms of motion of matter, there are also various forms of energy, which can be converted to each other, but the total amount remains the same. Energy can exist in numerous forms such as thermal, mechanical, kinetic, potential, electrical, magnetic and nuclear, and their sum constitutes the total energy of a system.

Quantitative attribute – the first law of thermodynamics: All kinds of energy are equal and have additive property, no distinction of quality.

Qualitative attribute – the second law of thermodynamics: The quality of energy can only go from high to low without outside help, and it is impossible to go from low to high spontaneously.

2.2 Introduction to the basic concept

This chapter focuses on the common basic concepts in engineering thermodynamics, such as heat engine, working medium, thermodynamic system, and boundary. Understanding and mastering these basic concepts are the basis of learning engineering thermodynamics. What's more, equilibrium state and state parameters, and equation of state will be introduced below. Followed by state parameter coordinate diagram, quasi-static process and reversible process will be discussed.

2.3 Thermodynamic system

2.3.1 Important concepts and definitions

(1) **Heat engine**: This is the general term for machines that can convert heat energy into mechanical energy. For example, steam engine, steam turbine, gas turbine, and internal combustion engine.
(2) **Working medium**: This is the medium that can realize the conversion between thermal energy and mechanical energy. Air, gas, and water vapor are common working media.
(3) **Heat source**: This has a large heat capacity. No significant change appears in its temperature and other thermodynamic parameters when it emits or absorbs a limited amount of heat.

https://doi.org/10.1515/9783111329703-002

(4) **Thermodynamic system**: In engineering thermodynamics, a certain working medium or space is usually chosen as the research object, which is called thermodynamic system, or system for short.
(5) **Surrounding or environment**: Objects outside the system.
(6) **Boundary**: This is the interface between the system and surrounding or environment.

As shown in Fig. 2.1, the dotted line represents the environment boundary. The system exchanges mass and heat with the environment across the boundary. Environment boundary can be fixed or mobile; meanwhile, it can be real or fictitious.

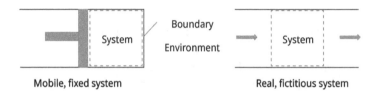

Mobile, fixed system   Real, fictitious system

Fig. 2.1: Thermodynamic system, environment, and boundary.

2.3.2 Classification of thermodynamic systems

There is no completely adiabatic or isolated system in nature, but there is a system close to adiabatic or isolated system in engineering. According to the situation in which a system interacts with the surrounding, system can be classified into closed system, open system, adiabatic system, and isolated system. Table 2.1 shows the specific interaction between the system and the surrounding, presenting whether there is mass or heat transfer.

Tab. 2.1: The interaction between the system and the surrounding.

	Yes	No
Whether the mass transfer	Open system	Closed system
Whether the heat transfer	Nonadiabatic system	Adiabatic system
Whether the work transfer	Nonwork isolation system	Work isolation system
Whether the mass, heat, work transfer	Nonisolated system	Isolated system

(1) **Closed system**: Closed system has no mass transfer with the surrounding. As we can see from Fig. 2.2, the mass of closed system remains constant, for which the closed system is also known as control mass system.
(2) **Open system**: Open system has mass transfer with the surrounding. As we can see from Fig. 2.2, there is mass exchange between the system and its surrounding

but no work transfer, for open system is a defined space. As a consequence, the open system is also known as a control volume system.
(3) **Adiabatic system**: Adiabatic system refers to the system that has no heat exchange with its surrounding.
(4) **Isolated system**: Isolated system has neither mass nor heat transfer with the surrounding.

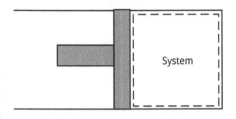

Closed system:
Mass is always constant, also known as **control mass system**

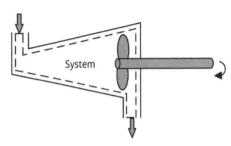

Open system:
A defined space, also known as **control volume system**

Fig. 2.2: Two typical thermodynamic systems.

2.4 Equilibrium state and state parameters

In the process of expansion or compression of working medium, pressure, temperature, volume, and other physical quantities will change accordingly. The macroscopic physical state presented by the working medium at a certain moment is called the thermal state of the working medium, or state for short. The macroscopic physical quantities used to describe the state of the working medium are called quantity of state, such as temperature, pressure, and specific volume.

2.4.1 Basic quantities of state

(1) **State (thermal state)**: This is the macroscopic physical state of a working medium at a certain moment.
(2) **Quantity of state**: A macroscopic physical quantity is used to describe the state of a working medium, such as temperature, pressure, and specific volume.
(3) **Equilibrium state**: This is the state in which, without external influence, the state parameters of a working medium (or system) do not change with time.
(4) **Non-equilibrium state**: When the temperature or pressure of each part in the system is inconsistent, there is energy or mass transfer between each part, and its quantity of state changes with time.

Characteristics of the quantity of state: When the state is determined, the quantities of state are also determined, and vice versa. Mathematically, it has the integral feature: the change of state parameters is independent of path, only related to initial and final states, as shown in Fig. 2.3.

No matter path 1-a-2 or path 1-b-2 is taken, the change of state parameters from state 1 to state 2 is the same, when the initial and final states are settled. Meanwhile, it means that the state parameters of circle 1-a-2-b-1 do not change, which is shown in the following equation:

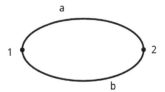

Fig. 2.3: Characteristics of the quantity of state.

$$\int_{1}^{2} dz = \int_{1,a}^{2} dz = \int_{1,b}^{2} dz = z_2 - z_1$$

$$\Rightarrow \oint dz = 0$$

(2.1)

2.4.2 Basic state parameters for description

In engineering thermodynamics, the common quantities of state are: temperature T, pressure p, specific volume v, thermodynamic energy U, enthalpy H, and entropy S.

Therein, temperature T, pressure p, and specific volume v can be directly measured experimentally, which are called basic state parameters.

(1) **Specific volume v:** The volume occupied by the working medium per unit mass, expressed by the symbol v, in m^3/kg.

(2) **Density ρ:** The mass of working medium per unit volume, expressed by the symbol ρ, in kg/m^3. Specific volume and density are reciprocal:

$$\rho v = 1$$

(2.2)

(3) **Pressure p:** Vertical force per unit area (i.e., pressure in physics), expressed by the symbol p, in Pa.

In the international system of units, pressure is measured in Pa, 1 Pa = 1 N/m². In engineering, kPa and MPa are often used as units of pressure. In addition, there are other commonly used pressure units, such as bar, mmH_2O, mmHg, atm (standard atmospheric pressure), and at (engineering atmospheric pressure). Their conversion relationship is shown in Tab. 2.2.

Tab. 2.2: The conversion relationship of pressure unit.

	Unit	Conversion relationship	
International system of units	Pa	1 kPa = 10^3 Pa	1 bar = 10^5 Pa
		1 MPa = 10^6 Pa	1 mbar = 10^2 Pa
Standard atmospheric pressure	atm	1 atm = 760 mmHg = 1.013×10^5 Pa	
		1 mmHg = 133.3 Pa	1 mmH$_2$O = 9.81 Pa
Engineering atmospheric pressure	at	1 at = 1 kgf/cm^2 = 9.80665×10^4 Pa	

In engineering, pressure gauge and U-shaped tube pressure gauge are often used to measure the pressure of working medium. Since the pressure gauge itself is always in an environment (normally atmospheric environment), the reading of the pressure gauge is usually the difference between the pressure of the working medium under test and the local ambient pressure, rather than the true pressure of the working medium. In addition, the true pressure of the working medium is the absolute pressure p and the local ambient pressure is the atmospheric pressure p_b.

When $p > p_b$, $p = p_b + p_g$, p_g is called the pressure gauge or gauge degree.

When $p < p_b$, $p = p_b - p_v$, p_v is called the vacuum gauge or vacuum degree.

In other words, gauge pressure $p_e = p - p_b$, and vacuum degree $p_v = p - p_b$.

(4) **Temperature T:** This is the physical quantity reflecting whether an object is hot or cold. And the temperature level reflects the violent degree of thermal movement of microscopic particles inside the object. The higher temperature means, the more intense the thermal motion of the molecules is.

Temperature is the criterion of thermal equilibrium. As shown in Fig. 2.4, if two systems are in thermal equilibrium with the third system, they must be in thermal equilibrium with each other, which is **the zeroth law of thermodynamics**. It is the establishment foundation of temperature concept and the theoretical basis of temperature measurement. Here is the origin of the zeroth law of thermodynamics by R. H. Fowler in 1939: *"When the laws of thermodynamics were originally established, there were only three. In the early eighteenth century, though, scientists realized that another law was needed to complete the set. However, this new law, which presented a formal definition of temperature, actually superseded the three existing laws and should rightfully be at the head of the list. This created a dilemma: the original three laws were already well known by their assigned numbers, and renumbering them would create a conflict with the existing literature and cause considerable confusion."*

The numerical representation of temperature is called temperature scale. The international system of units uses the thermodynamic temperature scale as the basic tempera-

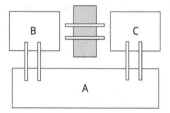

Fig. 2.4: The zeroth law of thermodynamics.

ture scale, and the temperature expressed by it is called the thermodynamic temperature (absolute temperature scale), expressed by the symbol T, and the unit is K (kelvin). For the thermodynamic temperature scale, the triple point of water is 273.16 K.

Meanwhile, there is the thermodynamic Celsius scale, or Celsius for short. The temperature expressed by it is called the Celsius temperature, expressed by the symbol t, and the unit is °C (degrees Celsius):

$$t = T - 273.15 \text{ K} \tag{2.3}$$

For the Celsius temperature scale, the triple point of water is 0.01 °C. And the temperature difference is 1 K = 1 °C.

2.5 Equation of state and state parameter coordinate diagram

The equilibrium state of a thermodynamic system can be described by state parameters. A thermal system has multiple state parameters, which describe some macroscopic characteristics of the system from different angles and are related to each other. As the state axioms said: for a simple compressible system, only two independent parameters are needed to determine its equilibrium state. When two of the three state parameters of the working medium in eq. (2.5) are determined, the last one is determined accordingly, such as eq. (2.4):

$$T = f(p, v) \tag{2.4}$$

$$pv = RT \tag{2.5}$$

The equation, which expresses the relationship between state parameters, is called the equation of state.

When there are two independent variables, they can be represented by a plan view. Any point on the plane represents an equilibrium state, and all states can be found on the plane, as shown in Fig. 2.5. State parameter coordinate diagram is a coordinate diagram with independent state parameters as coordinates and simple compression system as plane coordinates:

(1) Any equilibrium state of the system can be represented on the coordinate diagram
(2) Any point in the process line is an equilibrium state
(3) The non-equilibrium state cannot be represented in the diagram

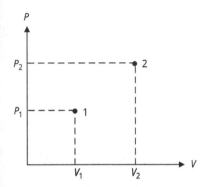

Fig. 2.5: Common p–v and T–s diagrams.

2.6 Quasi-static process and reversible process

2.6.1 Quasi-equilibrium process (quasi-static process)

Thermodynamic process is the change in the process of a system from one state to another. The change of state represents that the equilibrium of the system has been disturbed.

Due to the imbalance of temperature, pressure, or density in the system, the intermediate state experienced in the process of actual thermal equipment is unbalanced. If a process in which every state of the system is infinitely close to its equilibrium state, it is called quasi-equilibrium process or quasi-static process.

In the state parameter coordinate diagram, the quasi-equilibrium process can be approximately represented by continuous solid lines on the p–v diagram, such as p_1 and p_2 in Fig. 2.6.

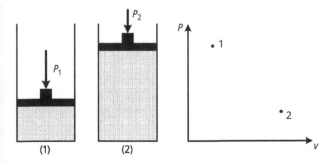

Fig. 2.6: Quasi-equilibrium process.

Here are some questions followed with answers which will help us to understand the basic concepts about different processes.

(1) Is it possible to change the state of the system and to get infinitely close to the equilibrium state at any time?
– Only by making the process infinitely slow.

(2) Is the actual process a balanced process?
– It is an unbalanced process with limited speed and time.

(3) When the actual process can be approximately seen as the quasi-equilibrium process?
– The unbalance potential (such as pressure difference and temperature difference) is not very large.
– The relaxation time (the time required to approach the equilibrium state from the non-equilibrium state) is very short.

The following examples illustrate the difference between a general process and a quasi-equilibrium process.

A typical general process is illustrated in Fig. 2.7:

$$p_1 = p_0 + \text{The weight} \tag{2.6}$$

$$T_1 = T_0 \tag{2.7}$$

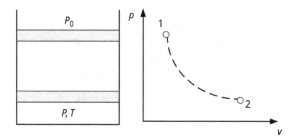

Fig. 2.7: The general process.

Suddenly remove the weight, finally we get:

$$p_2 = p_0 \tag{2.8}$$

$$T_2 = T_0 \tag{2.9}$$

A typical quasi-static process is illustrated in Fig. 2.8:

$$p_1 = p_0 + \text{The weight} \tag{2.10}$$
$$T_1 = T_0 \tag{2.11}$$

If the weight has infinite layers, remove only one infinite layer at a time; finally, the system is close to equilibrium at any time.

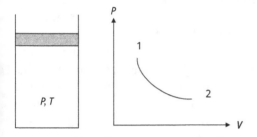

Fig. 2.8: The quasi-static process.

2.6.2 Engineering application of quasi-static process

Figure 2.9 illustrates a piston's internal combustion engine with 2,000 rpm, crank 2 stroke/turn, and 0.15 m/stroke:

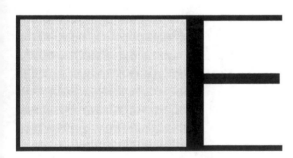

Fig. 2.9: Piston's internal combustion engine.

- Piston velocity: $2000 \times 2 \times 0.15/60 = 10$ m/s
- Pressure wave restoring equilibrium velocity (sound velocity): 350 m/s

Time to break balance (external action time) > time to restore balance (relaxation time)
    The general engineering process can be considered as a quasi-equilibrium process, but specific engineering problems need to be analyzed.

2.6.3 Reversible process

If the system completes a process and then reversely returns to its original state, leaving nothing to change, the process is called a reversible process. Otherwise, it is an irreversible process.

As shown in Fig. 2.10, we consider the working medium in the cylinder as the system. We start out at equilibrium 1. As the system absorbs heat from the heat source, the volume expands and pushes the piston to do work, driving the flywheel. The system goes through a series of quasi-equilibrium process from initial state 1 to final state 2.

If the device here is ideal, there is no friction loss, which means that all of the expansion work of the working medium is stored in the flywheel. If all the work of the flywheel is used to push the piston to compress the working medium under this ideal circumstance, the work of compression and impression between state 1 and state 2 is equal. As a consequence, when the system goes back to state 1, the device and the heat source also go back to their original state; in other words, the system and the surrounding both restored to their original states leaving no changes, which means the process is reversible.

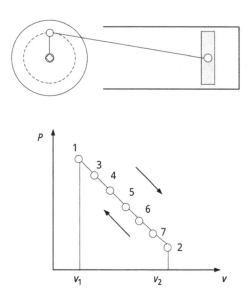

Fig. 2.10: Reversible process.

Therefore, we can come to a conclusion that a reversible process is an ideal process. Conditions for a reversible process = quasi-equilibrium process + no dissipation effect (friction, viscous disturbance, heat transfer with temperature difference, etc.).

The actual process is irreversible (such as heat transfer, mixing, diffusion, infiltration, dissolution, combustion, and electric heating)

2.7 Work and heat

2.7.1 Work and indicator diagram

2.7.1.1 Expansion work W

In mechanics, work is defined as the product of a force and the displacement along the direction in which the force acts. For example, if the body under the action of force F produces a small displacement dx in the direction of the force, then the amount of work done by the force is

$$\delta W = Fdx \tag{2.12}$$

If the body is displaced from x_1 to x_2 along the direction of the force under the action of force F, then the amount of work done by the change in force is

$$W = \int_{x_1}^{x_2} Fdx \tag{2.13}$$

The conversion of thermal energy into mechanical energy is achieved by volume expansion of the working medium. Work done by the working medium during volume expansion is called **expansion work** W with unit of J or kJ.

As shown in Fig. 2.11, we assume that the mass of gas in the cylinder is m, the pressure is p, and the piston area is A, which means that the force acting on the piston is pA.

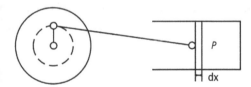

Fig. 2.11: Work done by the working medium during volume expansion.

Let us suppose that the piston is moved forward by a tiny distance dx by the pressure of the working medium. Since the expansion volume of the working medium is very small and its pressure is almost constant, this process can be regarded as a quasi-equilibrium process, then for the infinitesimal reversible process:

$$\delta W = pAdx = pdV \tag{2.14}$$

where dV is the displacement of the piston.

For the reversible process 1→2, as shown in Fig. 2.12, we can get:

$$W = \int_1^2 pdV \neq W_2 - W_1 \tag{2.15}$$

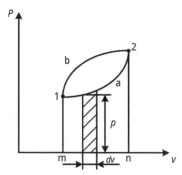

Fig. 2.12: The reversible process 1→2.

2.7.1.2 Specific expansion work w

Expansion work of unit mass working medium is called specific expansion work w with unit of J/kg or kJ/kg. For the expansion process, $dv > 0$ and $w > 0$. For the compression process, $dv < 0$ and $w < 0$:

$$\delta w = pdv \tag{2.16}$$

$$w = \int_1^2 pdv \tag{2.17}$$

The amount of w can be represented by the area under the process curve on the p–v diagram (Fig. 2.13).

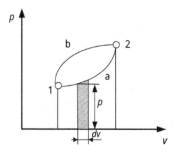

Fig. 2.13: Indicator diagram (p–v diagram).

2.7.1.3 Heat Q

The energy transferred by the temperature difference between the system and the outside world is called heat Q with unit of J or kJ. The heat transferred by unit mass of working medium is expressed by q, and its unit is J/kg or kJ/kg. The sign of heat is regulated as follows: for endothermic system, $q > 0$; for exothermic system, $q < 0$. Both heat and work are energy transferred during the interaction between the system and the outside world, which are process quantities rather than state quantities.

2.7.2 Heat, entropy, and thermogram

In the reversible process, the calculation formula of heat and work exchanged between the system and the outside world has the same form:

Work

$$\delta w = pdv \tag{2.18}$$

$$w = \int_1^2 pdv \tag{2.19}$$

Heat

$$\delta q = Tds \tag{2.20}$$

$$q = \int_1^2 Tds \tag{2.21}$$

According to the change of specific entropy, the direction of heat exchange between the system and the outside in a reversible process can be determined:
- $ds > 0$, $\delta q > 0$, system endothermic;
- $ds < 0$, $\delta q < 0$, system exothermic;
- $ds = 0$, $\delta q = 0$, system adiabatic; constant entropy process.

Thermogram
In the reversible process, the heat exchange between the unit mass of working medium and the outside world can be expressed by the area under the process curve on the T–S diagram (tephigram) which is shown in Fig. 2.14.

Here is a good example to understand how to calculate the work done in practical applications. There is a thermodynamic system with changing state: the relation between pressure and volume is $pV^{1.3}$ = constant, and the initial state of the thermodynamic system is p_1 = 600 kPa and V_1 = 0.3 m^3. If the system volume expands to V_2 = 0.5 m^3, please calculate the work done by the system to the environment.

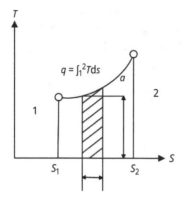

Fig. 2.14: Tephigram (also called thermogram).

Solution:

$$p = p_1 V_1^{1.3} / V^{1.3}$$

In this process, the work done by the system is

$$W = \int_{0.3}^{0.5} p dV = \int_{0.3}^{0.5} \frac{p_1 V_1^{1.3}}{V^{1.3}} dV = -\frac{p_1 V_1^{1.3}}{0.3} \left(V_2^{-0.3} - V_1^{-0.3} \right) = 85.25 \text{ kJ}$$

2.7.3 Thermodynamic cycle

This is the thermodynamic process in which a working medium starts from a certain initial state, goes through a series of thermodynamic states, and then returns to the original initial state. It is also known as a closed thermodynamic process, as shown in Fig. 2.15.

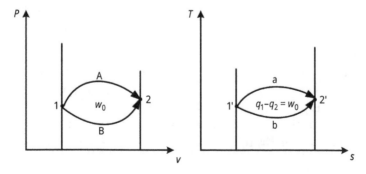

Fig. 2.15: Positive cycle (clockwise) and counter cycle (counterclockwise).

In Fig. 2.15, q_1 denotes the endothermic heat amount, q_2 denotes the exothermic heat amount, w_0 denotes the external work, and thus $w_0 = q_1 - q_2$. For the forward cycle, thermal efficiency $\eta_t = w_0/q_1$; for the reverse cycle, the refrigeration efficiency $\varepsilon = q_2/w_0$.

2.8 Summary

In this chapter, we have learned the basic concepts for thermodynamics and heat transfer sciences, including thermodynamic system, state (equilibrium/non-equilibrium), state parameter (such as p, v, T), equation of state (such as $pv = RT$), thermodynamic process (general process/quasi-static process, reversible process/irreversible process), work, heat, p–v diagram, and T–s diagram.

Question 1: What is the difference between equilibrium and steady?

Answer 1: We can respond to the above question as follows:
- Steady is when parameters do not change over time.
- Equilibrium is the absence of unbalanced potential difference.

There is a steady but non-equilibrium potential difference, as shown in Fig. 2.16.
- Steady is not necessarily an equilibrium.
- Equilibrium must be steady.

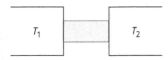

Fig. 2.16: The difference between equilibrium and steady.

Question 2: Is it right to say that an irreversible process is one that cannot be restored to its original state?

Answer 2: We can respond to the above question as follows:
- No, the key is whether the restoration to the initial state causes changes in the outside world.
- Reversible process means that if the system returns to its initial state, the outside world simultaneously returns to its initial state.
- Reversible does not mean that the system has to go back to its original state.

Question 3: What is the difference between a quasi-equilibrium process and a reversible process?

Answer 3: We can respond to the above question as follows:
- A reversible process must be a quasi-equilibrium process.
- A quasi-equilibrium process is not necessarily a reversible process.

Reversible process is an idealized concept, and then we always use the concept of a reversible process, but rarely use the concept of quasi-equilibrium process. Reversible process = quasi-equilibrium process + no dissipation effect.

Exercises

1. What is the difference between equilibrium state and stable state? Why is the concept of equilibrium state introduced into thermodynamics?
2. Can gauge pressure or vacuum be used as a state parameter for thermodynamic calculation? If the pressure of the working medium does not change, is it possible for the reading of the pressure gauge or vacuum gauge to change?
3. If the indicating value of the vacuum gauge is larger, will the actual pressure of the tested object be greater or smaller?
4. What is the difference between quasi-equilibrium process and reversible process?
5. Is it true that the irreversible process cannot be restored to its original state?
6. The cylindrical vessel shown in Fig. 2.17 has a diameter of 350 mm, the readings of pressure gauge A is 460 kPa, and pressure gauge B is 160 kPa. The atmospheric pressure is 100 mmHg. Try to obtain:
   (1) absolute pressure of vacuum chamber and 1 and 2 chambers;
   (2) reading of pressure gauge C; and
   (3) the force on the top of the cylinder.

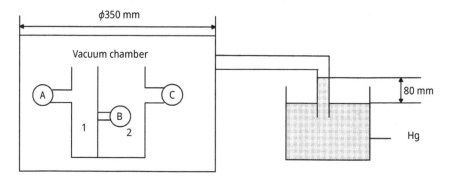

Fig. 2.17: Exercise 6 of additional graph.

7. The gas in a cylinder expands from 0.2 to 0.4 m³. If the pressure and volume of the gas are calculated in MPa and m³, respectively, the functional relationship between the pressure and volume of the gas in the expansion process is $p = 0.20V + 0.05$. Try to find the expansion work of gas.

Answers

1. The equilibrium state is a state in which the state parameters of the system do not change with time without being affected by the outside world. The stable state is a state in which the state parameters of the system do not change with time, no matter whether there is external influence or not. It can be seen that the equilibrium must be stable while stability is not necessarily in equilibrium. The concept of equilibrium state is introduced in thermodynamics to describe the macroscopic properties of the system with state parameters.

2. No. Because the gauge pressure or vacuum degree is a relative pressure, only absolute pressure can characterize the state of the working medium, which is the state parameter.

3. The smaller it is, the actual pressure is.

4. The quasi-equilibrium process without dissipation is a reversible process, so the reversible process must be a quasi-equilibrium process, while the quasi-equilibrium process is not necessarily a reversible process.

5. Incorrect. Irreversible process means that no matter with any complicated method, and the system cannot return to the initial state without any changes left by the outside world.

6. (1) $P_{va} = P_{at} - 80 \text{ mm Hg} = 100 \text{ mm Hg} - 80 \text{ mm Hg} = 20 \text{ mm Hg}$

   $P_1 = P_A + P_{va} = 460 \text{ kPa} + (20 \times 0.133) \text{ kPa} = 462.66 \text{ kPa}$

   $P_2 = P_1 - P_B = 462.66 - 160 = 302.66 \text{ kPa}$

   (2) $P_c = P_2 - P_{va} = 302.66 \text{ kPa} - (20 \times 0.133) \text{ kPa} = 300 \text{ kPa}$

   (3) $F = (P_b - P_{va}) \times (\pi/4 \times 0.35^2) = (100 - 20) \times 0.133 \text{ kPa} \times (\pi/4 \times 0.35^2) = 1{,}023.69 \text{ N}$

7. $W = \displaystyle\int_{0.2}^{0.4} (0.20V + 0.05) \times 10^6 \, dV = 2.2 \times 10^4 \text{ J}$

Chapter 3
First law of thermodynamics

3.1 Review: basic concepts

Basic concepts in engineering thermodynamics are the foundation of our study. Recall the definition of these basic concepts: heat engine, working medium, thermodynamic system, boundary, state, equilibrium state, specific volume, quasi-static process, and reversible process. Tell the differences and understand the essences of closed system, open system, adiabatic system, and isolated system.

3.2 Introduction to the first law of thermodynamics

The first law of thermodynamics is the law of energy conversion and conservation that all thermodynamic processes must follow. This chapter mainly focuses on the essence and mathematical description of the first law of thermodynamics, which is the theoretical foundation for calculation.

3.3 Storage energy of thermal system

The energy stored in the thermal system is called storage energy of the thermal **system**. There are mainly two types of stored energy of thermal system involved in engineering thermodynamics: one is thermodynamic energy that depends on the thermal system itself, and the other is the macroscopic kinetic energy related to the macroscopic displacement velocity of the system and the macroscopic potential energy related to the position of the system in the gravitational field.

3.3.1 Definition of thermodynamic energy

Thermodynamic energy is the sum of kinetic energy of thermal motion of molecules not involving chemical energy, atomic energy, and potential energy of interaction between molecules. And thermodynamic energy just depends on the thermal state of the system itself.

Thermal system storage energy:

$$E = U + E_k + E_p \tag{3.1}$$

https://doi.org/10.1515/9783111329703-003

Macroscopic kinetic energy:

$$E_k = mc^2/2 \qquad (3.2)$$

Macroscopic potential energy:

$$E_p = mgz \qquad (3.3)$$

Specific energy storage:

$$e = u + e_k + e_p \qquad (3.4)$$

Notes for thermodynamic energy (heat energy):
1. The thermodynamic energy of working medium with mass of m kg is expressed in U, and the unit is J or kJ.
2. The thermodynamic energy per unit mass (i.e., 1 kg) of working medium is called **specific thermodynamic energy**, which is expressed in u, and the unit is J/kg or kJ/kg.
3. The specific thermodynamic energy of gaseous working fluid only depends on the thermodynamic temperature and specific volume of the working fluid, that is, it depends on the thermodynamic state of the working fluid. Hence, it is a state parameter, which can be expressed as $u = f(T, v)$.
4. Thermodynamic energy is a state parameter, usually in the form of change dU, and its zero point can be specified artificially. For example, the thermodynamic energy of a gas is usually zero at 0 K.

3.4 The essence of the first law of thermodynamics

The first law of thermodynamics is essentially the law of energy conservation and transformation in the thermodynamic process; As shown in Fig. 3.1, the first law of thermodynamics can be described as follows: In the process of mutual conversion between thermal energy and other forms of energy, the total amount of energy is always the same.

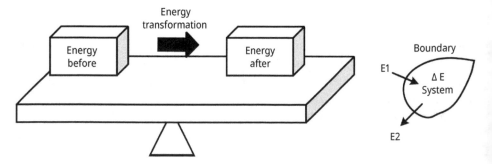

Fig. 3.1: The law of conservation.

The first type of perpetual motion machine

The first type of perpetual motion machine has two characteristics: one of that is it violates the first law of thermodynamics, and the other is that it can never do work without the input of energy forever.

One of the most famous design schemes of perpetual motion machine in the early stage was conceived by a Frenchman named Haneke in the thirteenth century. The discovery of the law of conservation of energy makes people realize that any machine can only transform energy from one form to another, rather than make energy out of nothing. So, **the first type of perpetual motion machine, which produces work without expending energy, is impossible to manufacture.**

3.5 The expression of the first law of thermodynamics for closed system

3.5.1 Fundamental thermodynamic relation

As shown in Fig. 3.2, we take the working medium in the cylinder as a closed system. Supposing that the heat absorbed from the surrounding when the system changes from equilibrium state 1 to equilibrium state 2 is Q and the expansion work is W, the mathematical expression of the system is:

For infinitesimal process:

$$Q = \Delta U + W \tag{3.5}$$

$$q = \Delta u + w \tag{3.6}$$

Suitable for any process and any working medium:

$$\delta Q = dU + \delta W \tag{3.7}$$

$$\delta q = du + \delta w \tag{3.8}$$

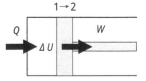

Fig. 3.2: Fundamental thermodynamic relation.

It reflects the conversion of heat and work. It shows that part of the heat added to the system is used to increase the thermodynamic energy of the working medium, and the rest is transferred to the outside world in the way of doing work and converted into mechanical energy.

3.5.2 Two expressions of the first law of thermodynamics for closed system

Example 1

As shown in Fig. 3.3, there is a refrigerator operating in a well-sealed and well-insulated room, whose door is open while the system is working.

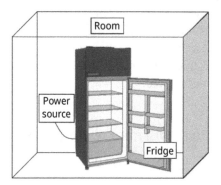

Fig. 3.3: Example 1: a refrigerator operating in a well-sealed and well-insulated room.

Method 1: select room as a research object
If we take the entire room including the air and the refrigerator as the system, then the room is a closed system:

$$Q = \Delta U + W \tag{3.9}$$

$$q = \Delta u + w \tag{3.10}$$

$$\Delta U + W = \Delta U - \left| W_{\text{electricity}} \right| = 0 \tag{3.11}$$

$$\Delta U = U_2 - U_1 = \left| W_{\text{electricity}} \right| > 0 \tag{3.12}$$

$$T_2 - T_1 > 0, \ T \uparrow \tag{3.13}$$

Method 2: select refrigerator as a research object
If we take the refrigerator as the system, the energy conversion diagram is shown in Fig. 3.4.

The refrigerator is a closed system:

$$\delta Q = dU + \delta W \tag{3.14}$$

$$\oint \delta Q = \oint dU + \oint \delta W \tag{3.15}$$

$$Q_{\text{net}} = W_{\text{net}} \tag{3.16}$$

$$|Q_1| - |Q_2| = -|W_{\text{electric}}| \tag{3.17}$$

$$|Q_2| - |Q_1| = |W_{\text{electric}}| > 0 \tag{3.18}$$

$$T_2 - T_1 > 0, T\uparrow \tag{3.19}$$

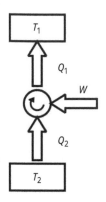

Fig. 3.4: Energy conversion diagram.

Example 2

As shown in Fig. 3.5, the working medium will expand freely in a confined space, when the partition is removed. In this circumstance, analyze ΔU.

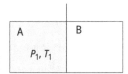

Fig. 3.5: Example 2: working medium expanding in a confined space.

Work and heat are the energy transferred through the boundary.

Take gas as a thermal system; this is a closed system:

$$Q = \Delta U + W \tag{3.20}$$

$$q = \Delta u + w \tag{3.21}$$

By the analysis of knowledge,

$$Q \to 0, \ W \to 0 \tag{3.22}$$

$$\Delta U = 0 \tag{3.23}$$

$$U_1 = U_2 \tag{3.24}$$

In summary:

For quasi-static process ($\delta w = pdv$):

$$Q = \Delta U + \int pdV \qquad (3.25)$$

$$\delta Q = dU + pdV \qquad (3.26)$$

$$q = \Delta u + \int pdv \qquad (3.27)$$

$$\delta q = du + pdv \qquad (3.28)$$

For reversible process ($\delta w = pdv$, $\delta q = Tds$)

$$\int TdS = \Delta U + \int pdV \qquad (3.29)$$

$$TdS = dU + pdV \qquad (3.30)$$

$$\int TdS = \Delta U + \int pdv \qquad (3.31)$$

$$Tds = du + pdv \qquad (3.32)$$

3.6 Stable flow energy equation of open system

3.6.1 Definition of stable flow

Stable flow is a flow that does not change over time, that is, the state of working medium on any flow section does not change with time, which is shown in Fig. 3.6.

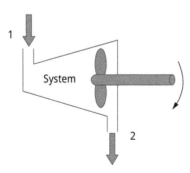

Fig. 3.6: Stable flow.

Realization conditions for stable flow:
1. The interaction of matter does not change with time
2. The interaction of energy does not change with time
3. The state parameters of import and export do not change with time

$$\dot{q}_{m1} = \dot{q}_{m2} = \text{const} \tag{3.33}$$

$$\frac{\delta q}{\delta \tau} = \text{const}, \quad \frac{\delta w}{\delta \tau} = \text{const} \tag{3.34}$$

3.6.2 Push work and flow work

3.6.2.1 Definition of push work
Push work is the energy transmitted by the working medium introduced or excluded by the system, which is shown in Fig. 3.7:

$$W_{\text{push}1} = \int_{0-0}^{1-1} pA\,dx = p_1 v_1 \tag{3.35}$$

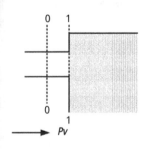 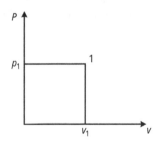

Fig. 3.7: Push work.

3.6.2.2 Definition of flow work
Flow work is the work required by the system to maintain flow. It is an energy added by a pump or fan to the conveyed working medium and transferred forward with the flow of the working medium, not the energy possessed by the working medium itself, which is shown in Fig. 3.8:

$$w_f = p_2 v_2 - p_1 v_1 = \Delta(pv) \tag{3.36}$$

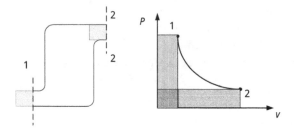

Fig. 3.8: Flow work.

3.6.3 Energy equation

3.6.3.1 Stable flow conditions: the first law of thermodynamics

Energy entering the system – energy leaving the system = changes in energy stored in the system

$$m_1 = m_2 = m \tag{3.37}$$

As shown in Fig. 3.9, we take the equipment as an open system and there are income working medium of mass m_1 and outcome working medium of mass m_2 in τ time.

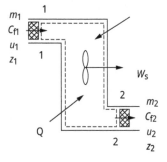

Fig. 3.9: Schematic diagram of opening system.

Energy entering the system in τ time:

$$Q + m\left(u_1 + \frac{1}{2}c_{f1}^2 + gz_1\right) + mp_1v_1 = Q + m\left(u_1 + \frac{1}{2}c_{f1}^2 + gz_1 + p_1v_1\right) \tag{3.38}$$

Energy leaving the system in τ time:

$$W_s + m\left(u_2 + \frac{1}{2}c_{f2}^2 + gz_2\right) + mp_2v_2 = W_s + m\left(u_2 + \frac{1}{2}c_{f2}^2 + gz_2 + p_2v_2\right) \tag{3.39}$$

3.6.3.2 Notes for thermodynamic energy (heat energy)

In the process of stable flow, the quantity and state of the working medium in the system do not change with time, so the total stored energy of the system remains unchanged.

According to the first law of thermodynamics:

$$\left[Q + mu_1 + p_1v_1 + \frac{1}{2}c_{f1}^2 + gz_1\right] - \left[W_s + mu_2 + p_2v_2 + \frac{1}{2}c_{f2}^2 + gz_2\right] = 0 \qquad (3.40)$$

Let $u + pv = h$, where h is called **enthalpy**, its physical meaning.

Specific enthalpy is a state parameter.

For a flow in a medium, the specific enthalpy represents the part of the total energy transferred forward along the flow direction per kilogram of the medium that depends on the thermal state.

The above formula can be sorted into:

$$Q = m\left(h_1 + \frac{1}{2}c_{f2}^2 + gz_2\right) - m\left(h_1 + \frac{1}{2}c_{f1}^2 + gz_1\right) + W_s \qquad (3.41)$$

$$Q = m\Delta h + \frac{1}{2}m\Delta c_f^2 + mg\Delta z + W_s \qquad (3.42)$$

Let $H = mh$, change the above formula to

$$Q = \Delta H + \frac{1}{2}m\Delta c_f^2 + mg\Delta z + W_s \qquad (3.43)$$

Therefore, the stable flow energy equation of an open system can be obtained.

For a unit mass working medium:

$$w = q - \Delta u = \Delta pv + \frac{1}{2}\Delta c_f^2 + g\Delta z + w_s \qquad (3.44)$$

For the infinitesimal process, the stable flow energy equation is written as

$$\delta Q = dH + \frac{1}{2}m dc_f^2 + mg dz + \delta W_s \qquad (3.45)$$

$$\delta q = dh + \frac{1}{2}dc_f^2 + g dz + \delta w_s \qquad (3.46)$$

Notes on specific enthalpy:

In engineering, only the change of enthalpy after the working medium that goes through a certain process needs to be calculated, rather than its absolute value, so the zero point of enthalpy value can be specified artificially.

The specific enthalpy is a state parameter for both flowing and nonflowing fluids.

For the flow medium, the flow work is equal to pv, and the specific enthalpy is the part of the total energy transferred forward along the flow direction by the unit mass medium, which depends on the thermal state.

For the nonflowing medium, there is no flow work, and the specific enthalpy does not represent the energy, but only the state parameter.

A work is required by the system to maintain flow. It is an energy added by a pump or fan to the conveyed working medium and transferred forward with the flow of the working medium, not the energy possessed by the working medium itself.

3.6.4 Technical work

In engineering thermodynamics, the sum of kinetic energy difference, potential energy difference, and shaft work that can be directly used in engineering technology is called **technical work**, which is expressed by W_t:

$$W_t = \frac{1}{2}m\Delta c_f^2 + mg\Delta z + W_s \tag{3.47}$$

For unit mass working medium:

$$w_t = \frac{1}{2}\Delta c_f^2 + g\Delta z + w_s \tag{3.48}$$

Stable flow energy equation of open system:

$$Q = \Delta H + W_t \tag{3.49}$$

$$q = \Delta h + w_t \tag{3.50}$$

Fundamental thermodynamic relation:

$$Q = \Delta U + W \tag{3.51}$$

$$q = \Delta u + w \tag{3.52}$$

For the infinitesimal process:

$$\delta Q = dH + \delta W_t \tag{3.53}$$

$$\delta q = dh + \delta w_t \tag{3.54}$$

Calculation method of technical work W_t:
According to the first law of thermodynamics, the expression of closed system is

$$q = \Delta u + w \tag{3.55}$$

Compared with the steady flow energy equation of the open system is:

$$\delta q = dh + \delta w_t \tag{3.56}$$

From the above analysis, the following can be obtained:

$$W_t = w - (\Delta h - \Delta u) = w - (p_2 v_2 - p_1 v_1) \tag{3.57}$$

Technical work = expansion work − net flow work
    When $p_2 v_2 = p_1 v_1$, technical work = expansion work
    For a reversible process:

$$w = \int_1^2 p\,dv \tag{3.58}$$

Substituting in the above formula:

$$W_t = \int_1^2 p\,dv - (p_2 v_2 - p_1 v_1) = \int_1^2 p\,dv - \int_1^2 d(pv) = -\int_1^2 v\,dp \tag{3.59}$$

where v is always positive; the negative sign represents that the sign of the technical work is opposite to that of dp. The process is reversible.

A graphical representation of technical work is shown in Fig. 3.10, where the area between the process curve and the y-coordinate represents the technical work of the reversible process.

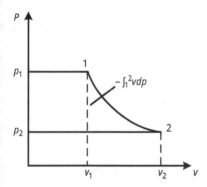

Fig. 3.10: Technical work.

Substituting the above equation into the stable flow energy equation of an open system:

$$q = \Delta h + w_t \text{ (applicable to general process)} \tag{3.60}$$

$$q = \Delta h - \int_1^2 v\,dp \text{ (applicable to reversible process)} \tag{3.61}$$

For the infinitesimal reversible process:

$$\delta q = dh - vdp \tag{3.62}$$

3.6.5 Unsteady flow process

Analysis and discussion of energy equation:

As shown in Fig. 3.11, an open system has energy inflows and outflows. In unsteady flow process, the general formula of energy equation of open system is

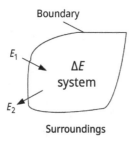

Boundary

E_1

ΔE
system

E_2

Surroundings

Fig. 3.11: An open system.

$$\delta Q = dE_{CV} + \left(h + \frac{1}{2}c_f^2 + gz \right)_2 \delta m_2 - \left(h + \frac{1}{2}c_f^2 + gz \right)_1 \delta m_1 + \delta W_s \tag{3.63}$$

3.7 The application of stable flow energy equation

3.7.1 Simplification of stable flow energy equation

Kinetic energy and potential energy changes are often neglected in thermodynamic problems. As a consequence, the stable flow energy equation can be simplified as: $q = \Delta h + w_s$.

In engineering, except for nozzle and diffuser, the kinetic energy and potential energy changes at the inlet and outlet of common thermal equipment are generally negligible.

3.7.2 Power machine and compressor

Power machines like steam turbine and gas turbine use the expansion of the working medium to do work and output shaft work w_s.

When working medium flows through fan, water pump, compressor, and other compression machinery, its pressure will rise, and the outside will exert shaft work on the working medium, which is exactly the opposite of the power machinery [8].

Due to the adoption of good heat preservation and insulation measures, the heat dissipation through the shell is very small, so it can be considered that the thermal process is an adiabatic process, that is, $q \approx 0$:

$$w_s = h_1 - h_2 \tag{3.64}$$

The output work is transformed by the enthalpy drop:

$$q = \Delta h + \frac{1}{2}\Delta c_f^2 + g\Delta z + w_s \tag{3.65}$$

$$w_s = -\Delta h = h_1 - h_2 \tag{3.66}$$

Compressor: $h_1 < h_2$, so $w_s < 0$

Power machine: $h_1 > h_2$, so $w_s > 0$

Figure 3.12 shows a land-based gas turbine, which is used for electric power generation. Gas turbine has core components of compressor, combustor, and turbine.

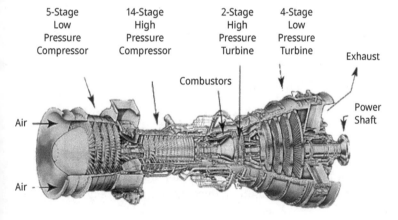

Fig. 3.12: A modern land-based gas turbine used for electric power production.

As shown in Fig. 3.13, the figure on the right is a steam power plant and the figure on the left is the schematic diagram, which includes boiler, superheater, steam turbine, electric generator, condenser, and feed water pump.

Figure 3.14 shows the simplified schematic diagram of a gas power plant. The air goes through compressor, being compressed. Then it will be mixed with fuel to go into gas turbine to do work, becoming exhaust gas and being discharged [9].

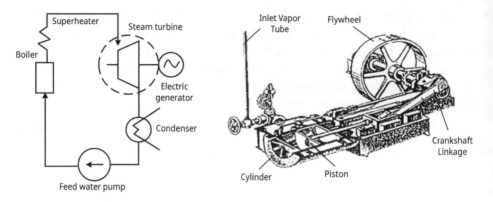

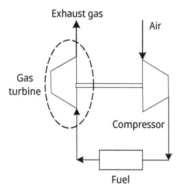

Fig. 3.13: Steam power plant.

Fig. 3.14: Gas power plant.

Figure 3.15 illustrates a schematic diagram of a refrigeration and air conditioning unit, which includes throttle valve, condenser, evaporator, and compressor, using the circulation and state change of refrigerant to reduce the temperature of the chamber and achieve refrigeration.

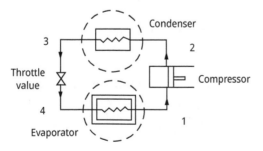

Fig. 3.15: Refrigeration and air conditioning unit.

3.7.3 Heat exchanger (heat exchange equipment)

Heat exchangers include heaters, radiators, evaporators, and condensers. As shown in Fig. 3.16, when the working medium flows through these devices, the reactive power is exchanged with the outside world, which means $w_s = 0$, and the change of kinetic energy and potential energy can be ignored:

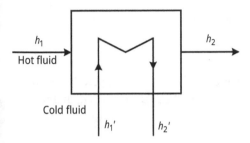

Fig. 3.16: Heat exchange equipment.

$$q = \Delta h + w_s \qquad (3.67)$$

No work units:

$$(w_s = 0) \quad q = \Delta h = h_2 - h_1 \qquad (3.69)$$

Heat release of hot fluid:

$$q = \Delta h = h_2 - h_1 < 0 \qquad (3.70)$$

Heat absorption of cold fluid:

$$q' = \Delta h = h_2' - h_1' > 0 \qquad (3.71)$$

3.7.3.1 Extended surface: tube–fin heat exchanger

Figure 3.17 shows the flow structure in a plain fin, where heat exchange surface is extended through adding fins and heat transfer is enhanced. The air flow is developing at the entry and wake zone emerges near the tube. There will be horseshoe vortices during the process.

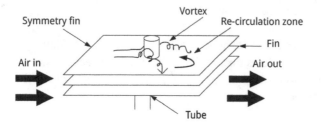

Fig. 3.17: Flow characteristics in a plain fin-and-tube heat exchanger.

3.7.3.2 Tubular: shell-and-tube heat exchanger

Figure 3.18 shows an annulus shell and helical coil tube heat exchanger. This heat exchanger is applied with a helical groove on the outer shell surface, strengthening the heat exchange.

Fig. 3.18: Shell-and-tube heat exchanger.

3.7.3.3 Plate type: plate heat exchanger

Figure 3.19 shows an exploded view of plate heat exchanger. There are corrugations on plates, increasing the plate surface area and enhancing heat transfer.

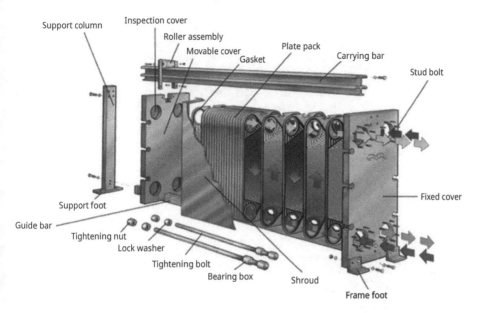

Support column
Inspection cover
Roller assembly
Movable cover
Gasket
Plate pack
Carrying bar
Stud bolt
Support foot
Guide bar
Tightening nut
Lock washer
Tightening bolt
Bearing box
Shroud
Frame foot
Fixed cover

Fig. 3.19: An exploded view of plate heat exchanger.

3.7.3.4 CPU radiator

Figure 3.20 shows a CPU radiator used for computer cooling, whose size is much smaller than common heat exchangers in our daily lives. It is the core of the computer, which affects the stability of the whole system.

Fig. 3.20: CPU radiator.

When the entire heat exchanger is selected as the control volume, q becomes zero, since the boundary for this case lies just beneath the insulation, and little or no heat crosses the boundary, which is shown in Fig. 3.21:

$$Q\overset{0}{\underset{\uparrow}{}} = \Delta H + \frac{1}{2}m\overset{0}{\underset{\uparrow}{}}\Delta c_f^2 + mg\overset{0}{\underset{\uparrow}{}}\Delta z + W_s\overset{0}{\underset{\uparrow}{}} \tag{3.72}$$

$$\Delta H = 0 \tag{3.73}$$

$$\Delta H_{hot} + \Delta H_{cold} = 0 \tag{3.74}$$

$$\left(H_2 + H_2'\right) - \left(H_1 + H_1'\right) = 0 \tag{3.75}$$

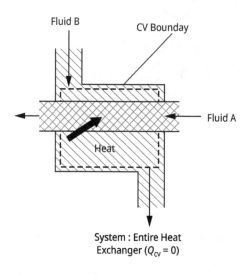

Fig. 3.21: Heat exchanger 1.

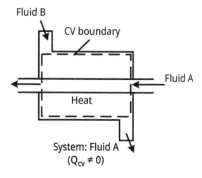

Fig. 3.22: Heat exchanger 2.

When only one of the fluids is selected as the control volume, then heat will cross the boundary as it flows from one fluid to the another and q will not be zero, which is shown in Fig. 3.22:

$$Q_{\uparrow}^{0} = \Delta h + \frac{1}{2}\Delta c_{f\,\uparrow}^{2\,0} + g_{\uparrow}^{0}\Delta z + W_{s\,\uparrow}^{0} \tag{3.76}$$

$$\Delta H = 0 \tag{3.77}$$

$$\Delta H_{hot} + \Delta H_{cold} = 0 \tag{3.78}$$

$$\left(H_2 + H_2'\right) - \left(H_1 + H_1'\right) = 0 \tag{3.79}$$

3.7.4 Adiabatic throttling

$$q_{\uparrow}^{0} = \Delta h + \frac{1}{2}\Delta c_{f\,\uparrow}^{2\,0} + g_{\uparrow}^{0}\Delta z + W_{s\,\uparrow}^{0} \tag{3.80}$$

$$\Delta h = h_2 - h_1 = 0 \tag{3.81}$$

As shown in Fig. 3.23, is throttling an isenthalpic process?

The adiabatic throttling process cannot be understood as a constant enthalpy process because of the large variation of flow velocity and the fact that the enthalpy is not equal everywhere between the two sections, especially near the necking.

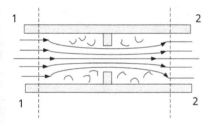

Fig. 3.23: A schematic diagram for adiabatic throttling process.

3.8 Summary

3.8.1 Essence

Application of law of conservation and transformation of energy in thermal phenomena.

Explain the energy conversion relationship of natural process from the perspective of energy quantity.

3.8.2 Energy equation

Closed system:

$$q = \Delta u + w \tag{3.82}$$

$$q = \Delta u + \int_1^2 p\,dv \tag{3.83}$$

Open system:

$$q = \Delta h + \frac{1}{2}\Delta c_f^2 + g\Delta z + w_s \tag{3.84}$$

$$q = \Delta u + w_t \tag{3.85}$$

$$q = \Delta u - \int_1^2 v\,dp \tag{3.86}$$

Eq. (3.82) applies to any working medium, any process; Eq. (3.83) applies to any working medium, reversible process; Eq. (3.84) applies to any working medium, any process; Eq. (3.85) applies to any working medium, any process; Eq. (3.86) applies to any working medium, reversible process.

3.8.3 Process quantity and state quantity

Types and laws of work
Expansion work W: caused by a change in volume, if reversible $w = \int p\,dv$.
Shaft work W_s: work output or input through the shaft of a turbomachinery.
Flow work W_f: the work required to maintain the flow $W_f = \Delta(pv)$.
Technical work W_t: technically usable work, if reversible $W_t = -\int v\,dp$

$$q - \Delta u = w = \Delta(pv) + \frac{1}{2}\Delta c_f^2 + g\Delta z + w_s \tag{3.87}$$

w is the expansion work, pv is the flow work, and $g\Delta z$ is the technical work.

Steps for energy analysis
1. Draw a diagram and select a thermal system.
2. Write the corresponding energy equation.
3. Analysis of interaction with the surroundings.
4. Simplify according to the specific situation.
5. Find the solution.

Question

1. What are the differences and relations among expansion work w, shaft work w_s, technical work w_t, and flow work w_f?

   Expansion work is the external work of the system due to volume change.

   Shaft work refers to the mechanical work exchanged between the thermal equipment and the outside when the working medium flows through the thermal equipment (open system), which is usually input and output through the shaft, so it is commonly called shaft work.

   The technical work includes not only the shaft work but also the change of mechanical energy (macrokinetic energy and potential energy) during the flow process of the working medium.

   Flow work is also called propulsion work. The flow work of 1 kg working medium is equal to the product of its pressure and specific volume. It is the work that the working medium transmits to the front of the flow, and only appears in the flow process of the working medium.

2. What are the differences and relations among expansion work w, shaft work w_s, technical work w_t, and flow work w_f?

$$q = \Delta u + w = \Delta h + w_t \tag{3.88}$$

$$h = u + pv \tag{3.89}$$

$$w = \Delta(pv) + w_t \tag{3.90}$$

Expansion work w = net flow work (Δw_f) + macrokinetic energy change + macropotential energy change (technical work w_t) + shaft work w_s.

Example

As shown in Fig. 3.24, air is compressed in a compressor. Before compression, the parameters of air are p_1 = 0.1 MPa, v_1 = 0.845 m³/kg. After compression, they are p_2 = 0.8 MPa, v_2 = 0.175 m³/kg. If the thermodynamic energy of each kg of air increases to 146.5 kJ in the process of compression, at the same time, 50 kJ will be released to the outside world; the compressor will produce 10 kg of compressed air per minute. Try to find out:

(1) The amount of compression work w done by per kilogram of air during compression.

(2) Shaft work w_s required to produce per kg of compressed air.

(3) How much power (kW) is required to drive this compressor?

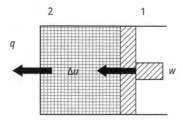

Fig. 3.24: A schematic diagram for air compressing process.

Solution

(1) The compression work per kg air during the compression process is

$$w = q - \Delta u = -50 - 146.5 = -196.5 \, \text{kJ/kg} \tag{3.91}$$

(2) Ignoring the changes in the macroscopic kinetic energy and potential energy of the gas inlet and outlet, the shaft work is equal to the technical work, so the shaft work required for the production of compressed air per kg is

$$w_s = q - \Delta u = q - (\Delta u + p_2 v_2 - p_1 v_1) = -50 - 146.5 - (0.8 \times 0.175 - 0.1 \times 0.845) \times 10^3$$

$$= -252 \, \text{kJ/kg}$$

(3) Therefore, the power required to drive this compressor is at least:

$$P = -\frac{w_s \times 10}{60} = 42 \, \text{kW} \tag{3.92}$$

Exercises

1. As shown in Fig. 3.25, in a closed system, working fluids change from state a to state b through the route a–c–b, absorbing 100 kJ heat, doing 40 kJ work outside. When through the route of a–d–b, doing 20 kJ work outside, try to calculate the heat exchange value between the working fluids and the environment through the route of a–d–b. When the working fluids change from state b to the initial state a through the curved line, the outside do 30 kJ work to the system, try to calculate the heat exchange value. If U_a = 0 kJ, U_d = 40 kJ, try to calculate the heat exchange value of processes a–d and d–b.

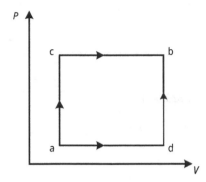

Fig. 3.25: Exercise 1: additional graph.

2. For a stable flow system, the gas parameters at the inlet are p_1 = 0.62 MPa, v_1 = 0.37 m³/kg, u_1 = 2,100 kJ/kg, c_{f1} = 300 m/s; the outlet parameters are p_2 = 0.13 MPa, v_2 = 1.2 m³/kg, u_2 = 1,500 kJ/kg, c_{f2} = 150 m/s. Mass flow rate of the gas q_m = 4 kg/s, and the heat exchanging outward through the system is 30 kJ/kg. Assume that the change of gravitational potential energy is ignored when the gas flows through the system. Try to calculate the output power of the gas flowing through the system.

3. There is a rigid adiabatic container, divided into two parts by a partition, with air on the left and a vacuum on the right.
   Questions:
   (1) How does the thermodynamic energy of air change?
   (2) Does the air do work?
   (3) Can this process be represented on a coordinate chart and why?

4. (1) Gas expansion will definitely do work to the outside world.
   (2) When the gas is compressed, it must consume the work from outside.
   (3) The gas must be heated to make it expand.
   (4) Simultaneous expansion and exotherm of the gas is possible.
   (5) It is impossible for gas to be compressed and suck heat at the same time.
   (6) It is impossible to heat the working medium, but its temperature decreases.

Answers

1. $\Delta U_{ab} = U_b - U_a = Q - W = 100 - 40 = 60 \text{ kJ}$

   $Q_{a-d-b} = \Delta U_{ab} + W = 60 + 20 = 80 \text{ kJ}$

   $Q_{b-a} = U_a - U_b + W = -\Delta U_{ab} + W = -60 - 30 = -90 \text{ kJ}$

   $\Delta U_{db} = U_b - U_d = 60 - 40 = 20 \text{ kJ}$

$$Q_{d-b} = U_d - U_b = \Delta U_{db} = 20 \text{ kJ}$$

$$Q_{a-d} = \Delta U_{ad} + W_{a-d} = \Delta U_{ad} + W_{a-d-b} = 40 + 20 = 60 \text{ kJ}$$

2. $$Q = \Delta H + \frac{1}{2} m \Delta c_f^2 + W_s$$

$$h_1 = u_1 + p_1 v_1 = 2,100 \times 10^3 + 0.62 \times 10^6 \times 0.37 = 2,329,400 \text{ J/kg}$$

$$h_2 = u_2 + p_2 v_2 = 1,500 \times 10^3 + 0.13 \times 10^6 \times 1.2 = 1,656,000 \text{ J/kg}$$

$$W_s = Q - \Delta H - \frac{1}{2} m \Delta c_f^2$$

$$= m \left(q - \Delta h - \frac{1}{2} \Delta c_f^2 \right)$$

$$= 4 \times \left[-30 \times 10^3 - (1,656,000 - 2,329,400) - \frac{1}{2} \times (150^2 - 300^2) \right]$$

$$= 4 \times 677.1 \text{ kW} = 2,708.6 \text{ kW}$$

3. (1) Thermodynamic energy remains unchanged.
    (2) Do no work externally.
    (3) No, because this is not a quasi-static process.
4. (1) False
    (2) True
    (3) False
    (4) True
    (5) False
    (6) False

Chapter 4
Property and process of the ideal gas

4.1 Review: first law of thermodynamics

Recall the essence and expression (for closed system) of the first law of thermodynamics, stable flow energy equation of open system, and the application of law of conservation and transformation of energy in thermal phenomena.

4.2 Introduction to property and process of the ideal gas

The ideal gas is a scientifically abstracted hypothetical gas that does not exist in nature. But in engineering, the properties of a gas working medium are close to ideal gas in most cases.

4.3 Ideal gas state equation

4.3.1 Ideal gas

Ideal gas–actual gas:
–   The ideal gas is a hypothetical gas, whose pressure is near zero and the volume is closer to infinity.
–   Many gases in engineering are close to the assumptions of the ideal gas when they are far from the liquid.
–   Ideal gas does not exist in nature, but some gas under a specific state can be approximated to the ideal gas:
    (1)   Actual gases such as O_2, N_2, H_2, CO under the pressure of 5 MPa and their mixtures at room temperature, such as air.
    (2)   Air with small amount of water vapor and low sub-pressure.

The working medium of a heat engine is usually made of gaseous substances: gas or steam.
   Gas: stay away from liquid, not easily to liquefied, such as air.
   Vapor: closer to liquid, easy to be liquefied, such as water vapor.

In the thermal properties, the relationship between p, v, and T is of particular importance. For actual gases, this relationship is generally complex. To this end, the ideal gas is proposed.

https://doi.org/10.1515/9783111329703-004

Macro definition:

If the basic state parameters of the gas p, v, and T satisfy, pv/T = const, it is called the ideal gas.

Micro definition:

Gas molecules are elastic and their mass is negligible; and there is no interaction between gas molecules.

4.3.2 Ideal gas state equation

According to the ideal gas macro definition:

$$\frac{pv}{T} = \text{const} \ggg pv = R_g T \tag{4.1}$$

where the units of p, v, T, and R_g are, respectively, Pa, m³/kg, K, and J/(kg · K). R_g is a gas constant, whose value depends on the type of gas, regardless of the state of the gas.

If the amount of the substance m is in kg, the amount of matter n is in mol, and M represents the mole mass of the substance:

$$n = \frac{m}{M} \tag{4.2}$$

The volume of 1 mol gas is indicated in V_m:

$$V_m = Mv \tag{4.3}$$

Both sides of eq. (4.1) are multiplied by M:

$$pMv = pV_m = MR_g T \tag{4.4}$$

1 mol gas under pressure p and temperature T with same molar volume:

$$V_{m1} = V_{m2} = V_{m3} = V_{m4} \tag{4.5}$$

MR_g is independent of the substance type and state, R become a mole gas constant. Hence:

$$R = MR_g = \frac{p_0 V_{m0}}{T_0} = \frac{101,325 \text{ Pa} \times 0.02241325 \text{ m}^3/\text{mol}}{273.15 \text{ K}} = 8.3143 \text{ J}/(\text{mol} \cdot \text{K}) \tag{4.6}$$

Four forms of expression

　　1 kg ideal gas (basic equation):

$$pv = R_g T \tag{4.7}$$

m kg ideal gas:

$$pV = mR_g T \tag{4.8}$$

1 mol ideal gas:

$$pV_m = RT \tag{4.9}$$

n mol ideal gas:

$$pV = nRT \tag{4.10}$$

Where:

$$V_m = Mv \tag{4.11}$$

$$R_g = \frac{R}{M} \tag{4.12}$$

$$n = \frac{m}{M} \tag{4.13}$$

4.4 Thermal capacity, thermodynamic energy, entropy, and enthalpy of ideal gases

4.4.1 Definition of thermal capacity

Heat capacity C:
Heat required to increase the temperature by 1 K (or 1 °C)

$$C = \frac{\delta Q}{dT} = \frac{\delta Q}{dt} \tag{4.14}$$

Specific heat capacity c:
The heat needed to increase 1 kg of substance by 1 K, and its unit is J/(kg · K)

$$c = \frac{\delta q}{dT} \tag{4.15}$$

Molar heat capacity C_m:
The heat needed to increase 1 mol of substance by 1 K, and its unit is J/(mol · K)

$$C_m = Mc \tag{4.16}$$

Factors affecting the heat capacity:
(1) Types of objects
(2) Mass of objects
(3) It is also related to the process, since heat is the amount of process

In thermal calculations, constant volume and constant pressure processes are often involved. Specific thermal capacity at constant volume c_V and specific thermal capacity at constant pressure c_p are two common thermal capacities.

Then derive their calculation formulas:

$$c_V = \frac{\delta q_V}{dT} = \left(\frac{\partial u}{\partial T}\right)_v \qquad (4.17)$$

$$c_p = \frac{\delta q_p}{dT} = \left(\frac{\partial h}{\partial T}\right)_p \qquad (4.18)$$

According to the first law of thermodynamics, the reversible process of microelements has the relationship:

$$\delta q = du + pdv \qquad (4.19)$$

Thermodynamic energy u is a state parameter, $u = u(T,v)$

$$du = \left(\frac{\partial u}{\partial T}\right)_V dT + \left(\frac{\partial u}{\partial v}\right)_T dv \qquad (4.20)$$

For constant volume, $dv = 0$

$$\delta q_V = \left(\frac{\partial u}{\partial T}\right)_V dT \qquad (4.21)$$

According to the definition formula:

$$c_V = \frac{\delta q_V}{dT} = \left(\frac{\partial u}{\partial T}\right)_v \qquad (4.22)$$

According to the first law of thermodynamics, the reversible process of microelements has the relationship:

$$\delta q = dh - vdp \qquad (4.23)$$

Thermodynamic energy u is a state parameter, $h = h(T, p)$

$$dh = \left(\frac{\partial h}{\partial T}\right)_p dT + \left(\frac{\partial h}{\partial p}\right)_T dp \qquad (4.24)$$

For constant pressure, $dp = 0$

$$\delta q_p = \left(\frac{\partial h}{\partial T}\right)_p dT \qquad (4.25)$$

According to the definition formula:

$$c_p = \frac{\delta q_p}{dT} = \left(\frac{\partial h}{\partial T}\right)_p \tag{4.26}$$

4.4.2 Specific heat capacity at constant pressure and specific heat capacity at constant volume of ideal gas

Thermodynamic energy contains internal kinetic energy and internal potential energy. There is no force between molecules for an ideal gas, so the ideal gas's thermodynamic energy only contains internal kinetic energy. The thermodynamic energy of the ideal gas is a single-value function of temperature T:

$$u = u(T) \tag{4.27}$$

$$h = u + pv = u + R_g T \tag{4.28}$$

– The enthalpy of the ideal gas is also a unit function of temperature: $h = h(T)$.

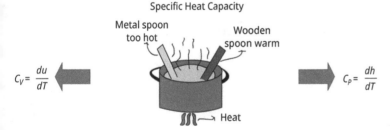

Fig. 4.1: Specific heat capacity.

As shown in Fig. 4.1, the relationship between c_p and c_v of ideal gas is

$$c_p = \frac{dh}{dT} = \frac{d(u+pv)}{dT} = \frac{du}{dT} + \frac{d(R_g T)}{dT} = c_V + R_g \tag{4.29}$$

Meyer formula:

$$c_p - c_V = R_g \tag{4.30}$$

Specific heat capacity ratio:

$$\gamma = \frac{c_p}{c_V} \tag{4.31}$$

$$c_p = \frac{\gamma}{\gamma - 1} R_g \tag{4.32}$$

$$c_v = \frac{1}{\gamma - 1} R_g \tag{4.33}$$

4.4.3 Calculation of the heat capacity of the ideal gas

Relationship between c_p and c_v of ideal gas is

$$C_p = a_0 + a_1 T + a_2 T^2 + a_3 T^3 + I \tag{4.34}$$

$$C_{pm} = a_0 + a_1 T + a_2 T^2 \tag{4.35}$$

a_0, a_1, a_2 are experimental fitting constants. Specific heat capacity at constant volume can also be obtained by the Meyer formula:

$$c_p = \frac{C_{pm}}{M} \tag{4.36}$$

$$c_v = \frac{C_{vm}}{M} \tag{4.37}$$

As shown in Fig. 4.2, for a constant pressure process and calculating the heat absorption:

$$q_p = \int_1^2 c_p(t)dt = c_p\Big|_{t_1}^{t_2}(t_2 - t_1) \tag{4.38}$$

$$c_p\Big|_{t_1}^{t_2} = \frac{\int_1^2 c_p(t)dt}{t_2 - t_1} \tag{4.39}$$

$$q_p = \int_1^2 c_p(t)dt = \int_{0°C}^{t_2} c_p(t)dt - \int_{0°C}^{t_1} c_p(t)dt = c_p\Big|_{0°C}^{t_2} \times t_2 - c_p\Big|_{0°C}^{t_1} \times t_1 \tag{4.40}$$

$$c_p\Big|_{t_1}^{t_2} = \frac{c_p\Big|_{0°C}^{t_2} \times t_2 - c_p\Big|_{0°C}^{t_1} \times t_1}{t_2 - t_1} \tag{4.41}$$

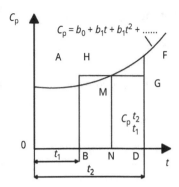

Fig. 4.2: c_p–t figure for a constant pressure process.

4.4.4 Linear relationship of average heat capacity

As shown in Fig. 4.3, for a constant volume process, calculate the heat absorption:

$$q_p = \int_1^2 c_p(t)dt = \int_1^2 (a + bt)dt = \left[a + \frac{b}{2}(t_1 + t_2)\right](t_2 - t_1) \tag{4.42}$$

$$c_p\Big|_{t_1}^{t_2} = a + \frac{b}{2}(t_1 + t_2) \tag{4.43}$$

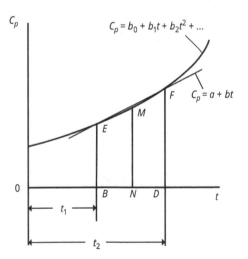

Fig. 4.3: c_p–t figure.

Constant molar heat capacity of ideal gas is

$$c_p = a \tag{4.44}$$

From molecular motion theory, thermodynamic energy of 1 mol ideal gas is

$$U_m = \frac{i}{2}RT\frac{1}{2} \qquad (4.45)$$

The specific heat capacity of constant volume heat capacity and constant pressure heat capacity of ideal gas can be obtained:

$$C_{V,m} = \frac{i}{2}R \qquad (4.46)$$

$$C_{p,m} = \frac{i+2}{2}R \qquad (4.47)$$

$$\gamma = \frac{i+2}{i} \qquad (4.48)$$

where i is the degree of freedom of molecular motion.

For single atom, $i = 3$; double atoms, $i = 5$; multiatoms, $i = 7$.

- The theory of motion of gas molecules and the principle of equal sharing of energy in terms of degrees of freedom indicate that gases with the same number of atoms have same heat capacity.
- Table 4.1 lists the molar heat capacities of single, double, and multiatom gases.

Tab. 4.1: The molar heat capacities of single, double, and multiatom gases.

	Single atom	Double atoms	Multiatoms
$C_{V,m}$	$\dfrac{3R}{2}$	$\dfrac{5R}{2}$	$\dfrac{7R}{2}$
$C_{p,m}$	$\dfrac{5R}{2}$	$\dfrac{7R}{2}$	$\dfrac{9R}{2}$
γ	1.67	1.4	1.29

4.5 Thermodynamics, entropy, and enthalpy of ideal gases

Relationship between c_p and c_v of ideal gas is given as follows:

Thermodynamic energy:

$$c_V = \frac{du}{dT} \qquad (4.49)$$

$$\Delta u = \int_1^2 c_V dT \qquad (4.50)$$

Enthalpy:

$$c_p = \frac{dh}{dT} \tag{4.51}$$

$$\Delta h = \int_1^2 c_p dT \tag{4.52}$$

Ideal gas is in either process: u and h; changes can be calculated by the above two ways.

Entropy of the ideal gas:

$$dS = \frac{\delta Q}{T} \tag{4.53}$$

Specific entropy:

$$ds = \frac{\delta q}{T} \tag{4.54}$$

In microelement reversible process, the increase in working medium entropy is equal to the heat absorbed by the working medium divided by the thermodynamic temperature of the working medium:

$$\delta q_{rev} = c_v dT + p dv \tag{4.55}$$

$$ds = \frac{c_v dT + p dv}{T} \tag{4.56}$$

$$\Delta s = \int_1^2 c_v \frac{dT}{T} + R_g \ln \frac{v_2}{v_1} \tag{4.57}$$

$$\delta q_{rev} = c_p dT - v dp \tag{4.58}$$

$$ds = \frac{c_p dT - v dp}{T} \tag{4.59}$$

$$\Delta s = \int_1^2 c_p \frac{dT}{T} - R_g \ln \frac{p_2}{p_1} \tag{4.60}$$

Entropy:
When the specific heat capacity is constant, entropy changes:

$$\Delta s = c_p \ln \frac{T_2}{T_1} - R_g \ln \frac{p_2}{p_1} \tag{4.61}$$

$$\Delta s = c_V \ln \frac{T_2}{T_1} + R_g \ln \frac{v_2}{v_1} \tag{4.62}$$

Differential equation of ideal gas:

$$\frac{dp}{p} + \frac{dv}{v} = \frac{dT}{T} \tag{4.63}$$

$$\Delta s = \int_1^2 c_p \frac{dT}{T} - R_g \ln \frac{p_2}{p_1} = \int_1^2 c_p \left(\frac{dp}{p} + \frac{dv}{v} \right) - R_g \ln \frac{p_2}{p_1} \tag{4.64}$$

$$\Delta s = c_V \ln \frac{p_2}{p_1} + c_p \ln \frac{v_2}{v_1} \tag{4.65}$$

Attention:

(1) The entropy change of ideal gas depends entirely on the initial and final states, so entropy change of the ideal gas is a state parameter.

(2) The upper formula is suitable to calculate the change of entropy of the ideal gas in any process.

4.6 Ideal gas mixture

4.6.1 Partial pressure and Dalton's law

Partial pressure p_i:

Each component individually occupies the same volume V as the mixed gas and the pressure at the temperature T of the mixed gas.

Dalton's law:

As shown in Fig. 4.4, the total pressure of the mixture gas p is equal to the sum of the partial pressure of each component.

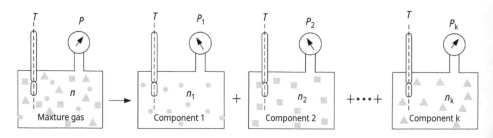

Fig. 4.4: Partial pressure and Dalton's law.

Quality conservation:

$$n = \sum n_i \tag{4.66}$$

$$m = \sum m_i \tag{4.67}$$

$$p_i V = m_i R_{gi} T \tag{4.68}$$

$$p_i V = n_i R T \tag{4.69}$$

General gas law:

$$\sum (p_i V) = \sum (n_i R T) \tag{4.70}$$

$$pV = nRT \tag{4.71}$$

4.6.2 Partial volume and partial volume law

Partial volume V_i:
Volume occupied by the ith component at the same pressure p and the same temperature T as the mixed gas.

Partial volume law:
As shown in Fig. 4.5, the total volume V of the mixed gas is equal to the sum of the component volumes.

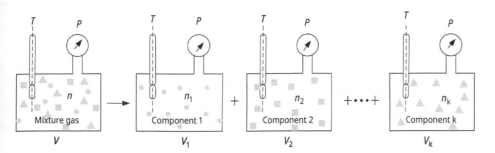

Fig. 4.5: Partial volume and partial volume law.

4.6.3 Ingredients of mixed gases

Mass fraction ω_i:
This is the ratio of the mass m_i and the total mass of mixed gas m:

$$\omega_i = \frac{m_i}{m}, \quad m = \sum_{i=1}^{k} m_i, \quad \sum_{i=1}^{k} \omega_i = 1 \tag{4.72}$$

Mole fraction x_i:

This is the ratio of the molar amount of meta-substance n_i and the amount of total mixture n:

$$x_i = \frac{n_i}{n}, \quad n = \sum_{i=1}^{k} n_i, \quad \sum_{i=1}^{k} x_i = 1 \tag{4.73}$$

Volume fraction φ_i:

This is the ratio of the volume of meta-substance V_i and the volume of total mixture V:

$$\phi_i = \frac{V_i}{V}, \quad V = \sum_{i=1}^{k} V_i, \quad \sum_{i=1}^{k} \phi_i = 1 \tag{4.74}$$

Relationship between ingredients is

$$\omega_i = \frac{m_i}{m} = \frac{n_i M_i}{\sum_{i=1}^{k} n_i M_i} = \frac{\frac{n_i}{n} M_i}{\sum_{i=1}^{k} \frac{n_i}{n} M_i} = \frac{x_i M_i}{\sum_{i=1}^{k} x_i M_i} \tag{4.75}$$

$$x_i = \frac{n_i}{n} = \frac{n_i}{\sum_{i=1}^{k} n_i} = \frac{m_i/M_i}{\sum_{i=1}^{k} (m_i/M_i)} = \frac{m_i/(m \cdot M_i)}{\sum_{i=1}^{k} m_i/(m \cdot M_i)} = \frac{\omega_i/M_i}{\sum_{i=1}^{k} \omega_i/M_i} \tag{4.76}$$

These relational formulas must be used under the premise of ideal gas.

4.6.3.1 Equivalent mole mass of the gas mixture and equivalent gas constant

State equation ideal gas mixture meets:

$$pV = nRT \tag{4.77}$$

For ideal gas mixture modeling on the pure mass of ideal gas:

$$m = \sum m_i = \sum n_i M_i = n M_{eq} \tag{4.78}$$

$$M_{eq} = \sum x_i M_i \tag{4.79}$$

where M_{eq} is the equivalent mole mass of the gas mixture.

Equivalent gas constant is

$$R_{g,eq} = \frac{R}{M_{eq}} \tag{4.80}$$

From the general gas law:

$$pV = \sum m_i R_{gi}\, T = m R_{g,eq} T \tag{4.81}$$

$$R_{g,eq} = \sum w_i R_{g,i} \tag{4.82}$$

$$m = \sum m_i = \sum n_i\, M_i = n M_{eq} \tag{4.83}$$

$$M_{eq} = \frac{m}{n} \tag{4.84}$$

The total mole amount of mixed ideal gas is equal to the sum of the amounts of each component:

$$n = n_1 + n_2 + I + n_k \tag{4.85}$$

$$\frac{m}{M_{eq}} = \frac{m_1}{M_1} + \frac{m_2}{M_2} + I + \frac{m_k}{M_k} \tag{4.86}$$

$$M_{eq} = \frac{1}{\frac{m_1}{m \cdot M_1} + \frac{m_2}{m \cdot M_2} + I + \frac{m_k}{m \cdot M_k}} = \frac{1}{\sum_{i=1}^{k} \frac{w_i}{M_i}} \tag{4.87}$$

$$m = m_1 + m_2 + I + m_k \tag{4.88}$$

$$nM = n_1 M_1 + n_2 M_2 + I + n_k M_k \tag{4.89}$$

$$R_{g,eq} = \frac{R}{M_{eq}} = R\sum_{i=1}^{k} \frac{w_i}{M_i} = \frac{R}{\sum_{i=1}^{k} x_i M_i} = \frac{R}{\sum_{i=1}^{k} \phi_i M_i} \tag{4.90}$$

4.6.4 Specific heat capacity of mixed gas

Total heat absorbed by the gas mixture is

$$Q = \sum Q_i \tag{4.91}$$

The heat absorbed by 1 kg mixture should be equal to the sum of the heat absorbed by each component:

$$q = \sum w_i q_i \tag{4.92}$$

According to the definition of specific heat capacity:

$$c = \frac{\delta q}{dT} = \sum \left(w_i \frac{\delta q_i}{dT} \right) = \sum w_i c_i \tag{4.93}$$

Specific constant pressure heat capacity and specific volume constant heat capacity:

$$c_p = \sum \omega_i c_{pi} \tag{4.94}$$

$$c_v = \sum \omega_i c_{vi} \tag{4.95}$$

4.6.4.1 Heat capacity of gas mixture

$$c_p = \sum_{i=1}^{k} \omega_i c_{p,i} \tag{4.96}$$

$$c_v = \sum_{i=1}^{k} \omega_i c_{v,i} \tag{4.97}$$

$$c_{p,m} = Mc_p = \sum_{i=1}^{k} x_i M_i c_{p,i} = \sum_{i=1}^{k} x_i c_{p,m,i} \tag{4.98}$$

$$c_{v,m} = Mc_v = \sum_{i=1}^{k} x_i M_i c_{v,i} = \sum_{i=1}^{k} x_i c_{v,m,i} \tag{4.99}$$

4.7 Purpose of thermal process of ideal gas

The purpose of studying the thermal process is to improve the efficiency of thermodynamic conversion in thermodynamic processes.
- Design: Depending on external conditions and operation process, form the best cycle
- Check: Thermal calculation of identified processes

Figure 4.6 shows some examples of thermal process in our daily lives.

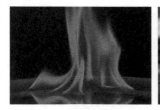 

Fig. 4.6: Examples of thermal process in our daily life.

- Studying the contents of the thermal process
- Change laws of parameters (p, T, v, u, h, s)
  Energy conversion relationship q, w, w_t

- Methods for studying thermal processes
- Abstract classification: constant volume, constant pressure, constant temperature, and constant entropy

Reversible process: If irreversible, do the re-correction
   Graphic method: p–v and T–s

4.7.1 The basis for studying the thermal process

Expression of the first law of thermodynamics:

$$q = \Delta u + w \tag{4.100}$$

$$q = \Delta h + w_t \tag{4.101}$$

Ideal gas state equation:

$$pv = R_g T \tag{4.102}$$

Characteristic relationship of reversible processes (w, w_t, q):

$$w = \int_1^2 p\,dv \tag{4.103}$$

$$w_t = -\int_1^2 v\,dp \tag{4.104}$$

$$q = \int_1^2 T\,ds \tag{4.105}$$

In actual situation, the actual process can be approximated to a typical process with simple laws:
- Constant volume process
- Constant pressure process
- Constant temperature process
- Thermal insulation process

4.7.2 The process of constant volume

4.7.2.1 Process in which the specific volume of gas remains constant
(1) Process equations: v = constant

(2) Relationship between initial and final parameters:

$$v_2 = v_1 \tag{4.106}$$

$$\frac{p_2}{p_1} = \frac{T_2}{T_1} \tag{4.107}$$

Changes of thermodynamic energy and enthalpy are:

$$\Delta u = \int_1^2 c_{vv} dT \tag{4.108}$$

$$\Delta h = \int_1^2 c_p dT \tag{4.109}$$

And more formulas are given in Tab. 4.2.

Tab. 4.2: The calculation formula for the various thermodynamic processes.

Process	Isobaric	Isometric	Isothermal	Adiabatic
Variable =>	Pressure	Volume	Temperature	No heat flow
Quantity Constant =>	$\Delta P = 0$	$\Delta V = 0$	$\Delta T = 0$	$Q = 0$
n	0	∞	1	$\gamma = C_p/C_v$
First law	$\Delta U = Q \cdot W$	$\Delta U = 0$ $W = 0$	$\Delta U = 0$ $Q = W$	$\Delta U = -W$ $Q = 0$
Work $W = \int PdW$	$P(V_2 - V_1)$	0	$P_1 V_1 \ln(V_2/V_1)$	$(P_1 V_1 - P_2 V_2)/(\gamma-1)$
Heat flow Q	$mC_p(T_2-T_1)$	$mC_v(T_2-T_1)$	$P_1 V_1 \ln(V_2/V_1)$	0
Heat capacity	C_p	C_v	∞	0
Internal energy $\Delta U = U_2 - U1$	$mC_p(T_2-T_1)$	$mC_v(T_2-T_1)$	0	$mC_v(T_2-T_1)$
Enthalpy $\Delta H = H_2 - H_1$ $H = U + PV$	$mC_v(T_2-T_1)$	$mC_p(T_2-T_1)$	0	$mC_p(T_2-T_1)$
Entropy $\Delta S = S_2 - S_1$ $= \int dQ/T$	$mC_p \ln(T_2/T_1)$	$mC_v \ln(T_2/T_1)$	$nR \ln(V_2/V_1)$	0^*
Ideal gas relations $P_1 V_1/T_1 = P_2 V_2/T_2$ $PV = NkT$	$P_1 = P_2$ $V_1/T_1 = V_2/T_2$ $T_1/T_2 = V_1/V_2$	$V_1 = V_2$ $P_1/T_1 = P_2/T_2$ $T_1/T_2 = P_1/P_2$	$T_1 = T_2$ $P_1 V_1 = P_2 V_2$ $P_1/P_2 = V_1/V_2$	$Q = 0$ $(S_1 = S_2)^*$ $P_1 V_1^\gamma = P_2 V_2^\gamma$ $T_1/T_2 = (V_2/V_1)^{\gamma-1}$

(3) Constant volume process in p–v and T–s figures

As shown in Fig. 4.7, the volume is constant, so the straight line perpendicular to the v-axis is

$$\delta q = du + pdv \tag{4.110}$$

For the process of constant volume:

$$\begin{cases} c_V = \frac{du}{dT} \\ ds = \frac{\delta q}{T} \end{cases} \tag{4.111}$$

$$ds = c_V \frac{dT}{T} \tag{4.112}$$

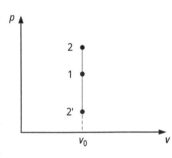

Fig. 4.7: p–v figure of process of constant volume.

If the specific heat capacity is set, integrating:

$$\int_{s_0}^{s} ds = \int_{T_0}^{T} c_V \frac{dT}{T} \tag{4.113}$$

$$T = T_0 e^{\frac{s-s_0}{c_{Vv}}} \tag{4.114}$$

(4) Representation of constant volume process in p–v and T–s figures

$$T = T_0 e^{\frac{s-s_0}{c_V}} \tag{4.115}$$

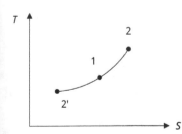

Fig. 4.8: T–s figure of process of constant pressure.

In Fig. 4.8, the constant volume process is the exponential function curve. The slope:

$$\left(\frac{\partial T}{\partial s}\right)_V = \frac{T}{c_V v} \tag{4.116}$$

The process is an exponential curve with a positive slope, since T and c_V are all above zero.

(5) Changes in initial and final thermodynamic energy, specific enthalpy, and specific entropy
Thermodynamic energy:

$$\Delta u = u_2 - u_1 = c_V(T_2 - T_1) \tag{4.117}$$

Specific enthalpy:

$$\Delta h = h_2 - h_1 = c_p(T_2 - T_1) \tag{4.118}$$

Specific entropy:

$$\Delta s = c_V \ln\frac{T_2}{T_1} + R_g \ln\frac{v_2}{v_1} = c_V \ln\frac{T_2}{T_1} \tag{4.119}$$

(6) The amount of work and heat in the process of constant volume
Because $dv = 0$, so expansion work is zero:

$$w = \int_1^2 p\,dv = 0 \tag{4.120}$$

Technical work

$$w_t = -\int_1^2 v\,dp = v(p_1 - p_2) \tag{4.121}$$

Heat

$$q = \Delta u \tag{4.122}$$

4.7.3 The process of constant pressure

1. Process equations: p = constant

2. Relationship between initial and final parameters

As shown in Fig. 4.9, the pressure is constant, so the straight line is perpendicular to the p-axis:

$$p_2 = p_1 \tag{4.123}$$

$$\frac{v_2}{v_1} = \frac{T_2}{T_1} \tag{4.124}$$

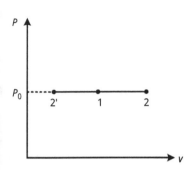

Fig. 4.9: p–v figure of constant pressure.

3. Representation of constant pressure process in p–v and T–s figures
Constant pressure process is a straight line perpendicular to the p-axis.
    For the process,

$$ds = c_p \frac{dT}{T} \tag{4.125}$$

If heat capacity is set, integrating, we can get:

$$T = T_0 e^{\frac{s-s_0}{c_p}} \tag{4.126}$$

In T–S figure of Fig. 4.10, the constant pressure process is also an exponential function curve:

$$\left(\frac{\partial T}{\partial s}\right)_p = \frac{T}{c_p} < \left(\frac{\partial T}{\partial s}\right)_V = \frac{T}{c_V} \tag{4.127}$$

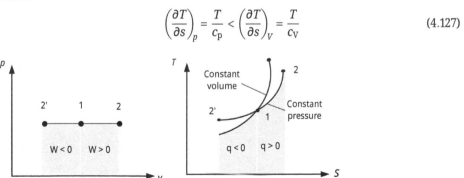

Fig. 4.10: Constant pressure process: p–v and T–S figures.

(1) Changes in initial and final thermodynamic energy, specific enthalpy, and specific entropy

Thermodynamic energy:

$$\Delta u = u_2 - u_1 = c_v(T_2 - T_1) \tag{4.128}$$

Specific enthalpy:

$$\Delta h = h_2 - h_1 = c_p(T_2 - T_1) \tag{4.129}$$

Specific entropy:

$$\Delta s = c_p \ln\frac{T_2}{T_1} - R_g \ln\frac{p_2}{p_1} = c_v \ln\frac{T_2}{T_1} \tag{4.130}$$

(2) The amount of work and heat in the process of constant pressure

Expansion work:

$$w = \int_1^2 p\,dv = p(v_2 - v_1) \tag{4.131}$$

Technical work:

$$w_t = -\int_1^2 v\,dp = 0 \tag{4.132}$$

Heat:

$$q = \Delta h = h_2 - h_1 \tag{4.133}$$

4.7.4 The process of constant temperature

The process that gas temperature remains the same:
- Process equation: T = constant, $pv = R_g T$, pv = constant
- Relationship between initial and final parameters: $T_2 = T_1$

$$\frac{p_2}{p_1} = \frac{v_1}{v_2} \tag{4.134}$$

Representation of constant temperature process in p–v and T–s figures

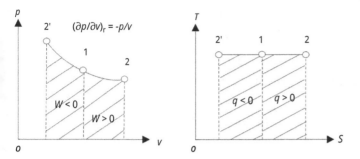

Fig. 4.11: Constant temperature process: p–v and T–s figures.

In p–v figure of Fig. 4.11, constant temperature process is an equilateral hyperbola line.

Changes in initial and final thermodynamic energy, specific enthalpy, and specific entropy:

Thermodynamic energy:

$$\Delta u = u_2 - u_1 = c_v(T_2 - T_1) = 0 \tag{4.135}$$

Specific enthalpy:

$$\Delta h = h_2 - h_1 = c_p(T_2 - T_1) = 0 \tag{4.136}$$

Specific entropy:

$$\Delta s = c_p \ln \frac{T_2}{T_1} - R_g \ln \frac{p_2}{p_1} = R_g \ln \frac{v_2}{v_1} \tag{4.137}$$

The amount of work and heat in the process of constant temperature:
Expansion work:

$$w = \int_1^2 p\,dv = \int_1^2 \frac{R_g T}{v}\,dv = R_g T \ln \frac{v_2}{v_1} = R_g T \ln \frac{p_1}{p_2} \tag{4.138}$$

Technical work:

$$w = -\int_1^2 v\,dp = -\int_1^2 \frac{R_g T}{p}\,dp = R_g T \ln \frac{p_1}{p_2} = w \tag{4.139}$$

Constant temperature process for ideal gas, $\Delta u = \Delta h = 0$.

According to the expression of the law of thermodynamics,

$$q = w = w_t = R_g T \ln \frac{v_2}{v_1} = R_g T \ln \frac{p_1}{p_2} \qquad (4.140)$$

Heat can also be calculated by changes in entropy:

$$q = \int_1^2 T ds = T(s_2 - s_1) \qquad (4.141)$$

This is applicable to the constant temperature process of actual gas or liquid.

4.7.5 Adiabatic process

There is no heat exchange between the gas and the outside world ($q = 0$).
For the reversible adiabatic,

$$ds = \frac{\delta q}{T} = 0 \qquad (4.142)$$

Reversible + adiabatic process = the process of constant entropy

(1) Process equation $pv^k = $ const
κ is the adiabatic index, equal to the ratio of heat capacity γ, $\kappa > 1$

(2) Relationship between initial and final parameters:

$$p_1 v_1{}^k = p_2 v_2{}^k \qquad (4.143)$$

$$\frac{p_2}{p_1} = \left(\frac{v_1}{v_2}\right)^k \qquad (4.144)$$

$$pv = R_g T \qquad (4.145)$$

The above formula can be converted into:

$$\frac{T_2}{T_1} = \left(\frac{v_1}{v_2}\right)^{k-1} \qquad (4.146)$$

$$\frac{T_2}{T_1} = \left(\frac{p_2}{p_1}\right)^{\frac{k-1}{k}} \qquad (4.147)$$

(3) Representation of adiabatic process in p–v and T–s figures
As shown in the T–s figure of Fig. 4.12, adiabatic process is a straight line perpendicular to the s-axis:

$$\frac{T_2}{T_1} = \left(\frac{p_2}{p_1}\right)^{\frac{n-1}{n}} \tag{4.159}$$

Δu, Δh, and Δs: Calculated according to the formula for the ideal gas:

$$\Delta u = \int_1^2 c_v dT \tag{4.160}$$

$$\Delta h = \int_1^2 c_p dT \tag{4.161}$$

$$\Delta s = c_v \ln \frac{p_2}{p_1} + c_p \ln \frac{v_2}{v_1} \tag{4.162}$$

(3) Representation of adiabatic process in p–v and T–s figures:

$$pv^n = \text{constant} \tag{4.163}$$

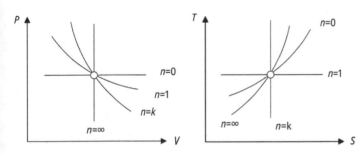

Fig. 4.14: P–v and T–s figures.

Distribution of polytropic process lines is shown in Fig. 4.14.
    Positive and negative judgment of Δu, Δh, w, w_t, q in the process:
    Δu, Δh, and dT are of the same sign

$$\Delta u = \int_1^2 c_v dT \tag{4.164}$$

$$\Delta h = \int_1^2 c_p dT \tag{4.165}$$

As shown in Fig. 4.15, when the process line is above the line of $n = 1$, dT is positive, which means that Δu and Δh are positive:

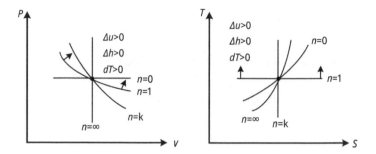

Fig. 4.15: dT, Δu, and Δh analysis in P–v and T–s figures.

$$w = \int_1^2 pdv \tag{4.166}$$

As shown in Fig. 4.16, when the process line is on the right side of the line of $n = \infty$, dv is positive, which means that w is positive.

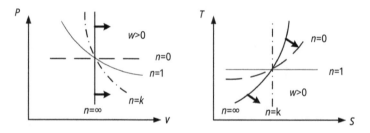

Fig. 4.16: w analysis in P–v and T–s figures.

(4) Representation of adiabatic process in P–v and T–s figures:
Positive and negative judgment of Δu, Δh, w, w_t, and q in the process

$$w_t = -\int_1^2 v\, dp \tag{4.167}$$

As shown in Fig. 4.17, when the process line is below the line of $n = 0$, dp is positive, which means that w_t is positive:

$$q = \int_1^2 T\, ds \tag{4.168}$$

As shown in Fig. 4.18, when the process line is above the line of $n = 1$, ds is positive, which means that q is positive.

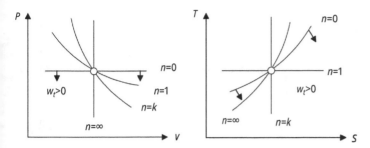

Fig. 4.17: w_t analysis in P–v and T–s figures.

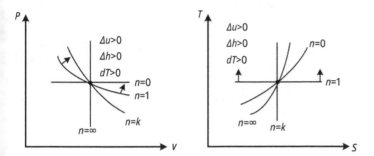

Fig. 4.18: q analysis in P–v and T–s figures.

Amount of work and heat in the polytropic process:

For expansion work w:

$$w = \int_1^2 p \, dv \tag{4.169}$$

where

$$p = p_1 v_1^n / v^n \tag{4.170}$$

$$pv = R_g T \tag{4.171}$$

$$\frac{T_2}{T_1} = \left(\frac{p_2}{p_1}\right)^{\frac{n-1}{n}} \tag{4.172}$$

which means

$$w = \int_1^2 p \, dv = \frac{1}{n-1}(p_1 v_1 - p_2 v_2)$$

$$= \frac{1}{n-1} R g T_1 \left[1 - \left(\frac{p_2}{p_1}\right)^{\frac{n-1}{n}}\right] = \frac{1}{n-1} R g (T_1 - T_2) \tag{4.173}$$

where $n \neq 0, 1$.

For technical work w_t:

$$w_t = -\int_1^2 v\, dp = n \int_1^2 p\, dv = n \cdot w \tag{4.174}$$

$$w = \int_1^2 p\, dv \tag{4.175}$$

where

$$pv^n = \text{const} \tag{4.176}$$

which means

$$vdp = -npdv \tag{4.177}$$

$$w_t = -\int_1^2 vdp = n \int_1^2 pdv = n \cdot w \tag{4.178}$$

where $n \neq \infty$.

For heat:

When $n = 1$, the process is of constant temperature, $\Delta u = 0$:

$$q = \Delta u + w = w \tag{4.179}$$

When $n \neq 1$, if the specific capacity is assumed constant:

$$
\begin{aligned}
q = \Delta u + w &= C_v(T_2 - T_1) + R_g(T_1 - T_2) \\
&= \left(C_v - \frac{R_g}{n-1} \right)(T_2 - T_1) \\
&= \left(\frac{R_g}{\kappa - 1} - \frac{R_g}{n-1} \right)(T_2 - T_1)
\end{aligned}
\tag{4.180}
$$

and

$$\text{polytropic capacity } C_n = \frac{n - \kappa}{n - 1} C_v \tag{4.181}$$

$$q = \frac{n - \kappa}{n - 1} C_v(T_2 - T_1) = C_n(T_2 - T_1) \tag{4.182}$$

(5) Polytropic process

$$C_n = \frac{n - \kappa}{n - 1} C_v \tag{4.183}$$

When there is constant pressure, $n = 0$, $C_n = C_p$.
When there is constant temperature, $n = 1$, $C_n \to \infty$.
When there is adiabatic process, $n = \kappa$, $C_n \to 0$.
When there is constant volume, $n \to \infty$, $C_n = C_v$

In order to facilitate comparison, the formulas of four typical thermal processes and polytropic processes are summarized in Tab. 4.3.

Tab. 4.3: Calculation formulas for various thermal processes.

Process	Relationship between p, V, T	Work (W)	Internal energy change (U_2-U_1)	Heat (Q)
Constant pressure	$p = \text{constant}$ $V_1/T_1 = V_2/T_2$	$p(V_2-V_1)$	$mC_v(T_2-T_1)$	$mC_p(T_2-T_1)$
Constant volume	$V = \text{constant}$ $V_1/T_1 = V_2/T_2$	0	$mC_v(T_2-T_1)$	$Q = U_2-U_1$ $Q = mC_v(T_2-T_1)$
Isothermal	$T = C$ $p_1V_1 = p_2V_2$	$p_1V_1 \ln(V_2/V_1)$	0	$Q = W$ $Q = p_1V_1 \ln(V_2/V_1)$
Polytropic	$pV^n = \text{constant}$ $T_2/T_1 = (V_1/V_2)^{n-1}$ $p2/p1 = (T_2/T_1)^{n/n-1}$	$(p_1V_1 - p_2V_2)/(n-1)$	$mC_v(T_2-T_1)$	$Q = W + U_2-U_1$
Adiabatic	As for polytropic with $n = \gamma$	As for polytropic with $n = \gamma$	$mC_v(T_2-T_1)$	0

4.8 Summary

Focus on the following:
(1) Calculate the ideal gas state parameter using the ideal gas state equation:

$$pv = R_g T$$

(2) Master the definition of heat capacity and the characteristics of ideal gas heat capacity:

$$c = \frac{\delta q}{dT} = \frac{\delta q}{dt} \tag{4.184}$$

and

$$c_p = \frac{\delta q_p}{dT} = \left(\frac{\partial h}{\partial T}\right)_p \tag{4.185}$$

and

$$c_v = \frac{\delta q_v}{dT} = \left(\frac{\partial h}{\partial T}\right)_v \qquad (4.186)$$

Gets:

$$c_p = \frac{dh}{dT} \qquad (4.187)$$

and

$$c_v = \frac{du}{dT} \qquad (4.188)$$

$$c_p - c_v = R_g \qquad (4.189)$$

(3) Master the calculation method of ideal gas thermodynamic energy, enthalpy, and entropy:

$$du = c_v dT \qquad (4.190)$$

$$\Delta u = \int_1^2 c_v dT \qquad (4.191)$$

and

$$dh = c_p dT \qquad (4.192)$$

$$\Delta h = \int_1^2 c_p dT \qquad (4.193)$$

$$\Delta s = c_v \ln\frac{p_2}{p_1} + c_p \ln\frac{v_2}{v_1} \qquad (4.194)$$

(4) Master the process equation of the basic thermodynamic process of the ideal gas, the change of the state parameters, the representation on the state parameter coordinate graph, and the calculation method of the amount of work and heat in the process. As shown in Fig. 4.19.

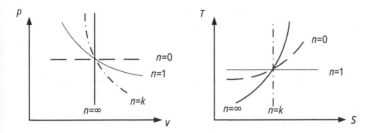

Fig. 4.19: *P–v* and *T–s* figures.

Exercises

1. An adiabatic rigid body cylinder is divided into two parts by a thermally conductive, frictionless piston. As shown in Fig. 4.20. Initially, the piston is fixed in a certain position. One side of the cylinder stores 0.4 MPa of ideal gas at 30 °C and 0.5 kg, and the other side stores 0.12 MPa, 30 °C, 0.5 kg of the same gas, then relax the piston to allow it to move freely, and the last two sides reach the equilibrium. The provided specific heat capacity is constant, and the test requirements are: (1) temperature (°C) at equilibrium and (2) the equilibrium pressure (MPa).

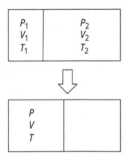

Fig. 4.20: Exercise 1: additional graph.

2. Present the polytropic process meeting following requirements in the *p–v* and *T–s*: air, boost pressure and temperature, and release heat. (The answer is shown in the Fig. 4.21)

3. Present the polytropic process meeting following requirements in the *p–v* and *T–s*: Expansion process with $n = 1.6$, and judge the sign of q, w, and Δu. (The answer is shown in the Fig. 4.22)

4. 6 kg of air, expanded from the initial state $p_1 = 0.3$ MPa, $t_1 = 30$ °C, to the final pressure $p_2 = 0.1$ MPa after a constant entropy process. Try to find the expansion work, heat exchanged, and final temperature of the air during this process.

5. Draw a reversible constant volume heating process, a reversible constant pressure heating process, a reversible constant temperature heating process, and a reversible adiabatic expansion process on a p–v diagram. (The answer is shown in the Fig. 4.23)

6. Show the polytropic process that meets the following requirements of air on the p–v diagram and T–s diagram:

   (1) The air pressure boosts, heats up, and releases heat (The answer is shown in the Fig. 4.24)

   (2) The air expands, heats up, and absorbs heat (The answer is shown in the Fig. 4.25)

   (3) Expansion process with $n = 1.6$, judge the sign of q, w, Δu (The answer is shown in the Fig. 4.26)

   (4) Expansion process with $n = 1.3$, judge the sign of q, w, Δu (The answer is shown in the Fig. 4.27)

Answers

1. (1) The temperature at equilibrium is 30 °C.

   (2) $p_1V_1 = \dfrac{m_1}{M}RT_1 \rightarrow V_1 = \dfrac{m_1}{MP_1}RT_1$

   Similarly: $V_2 = \dfrac{m_2}{Mp_2}RT_2$ $V = \dfrac{m}{Mp}RT$

   $$V_1 + V_2 = V \rightarrow \dfrac{m_1}{MP_1}RT_1 + \dfrac{m_2}{Mp_2}RT_2 = \dfrac{m}{Mp}RT$$

   Finishing: $\dfrac{m_1}{P_1}T_1 + \dfrac{m_2}{p_2}T_2 = \dfrac{m}{p}T$

   Get: $p = 0.185$ MPa.

2.

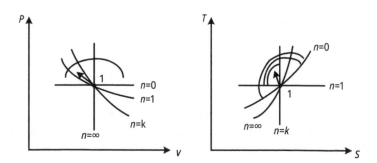

Fig. 4.21: Exercise 2: answer.

3.

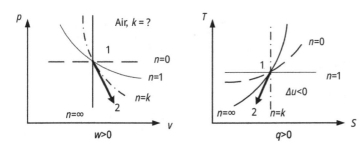

Fig. 4.22: Exercise 3: answer.

4.

$$T_2 = T_1 \left(\frac{p_2}{p_1}\right)^{\frac{k-1}{k}} = 303 \times \left(\frac{0.1}{0.3}\right)^{\frac{1.4-1}{1.4}} = 221 \text{ K}$$

$$W = m \frac{Rg}{k-1}(T_1 - T_2) = 6 \times \frac{287}{1.4-1}(303 - 221) = 352 \text{ kJ}, \ Q = 0$$

5.

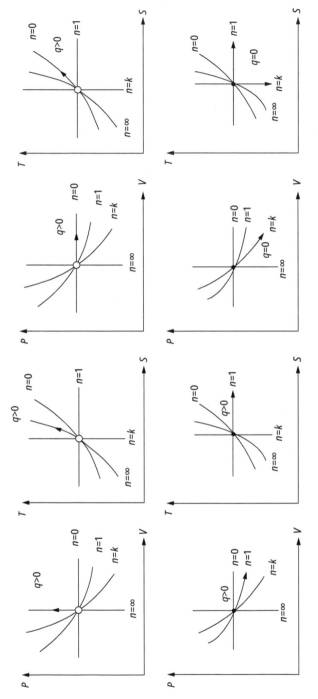

Fig. 4.23: Exercise 5: answer.

6. (1)

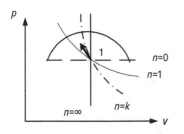

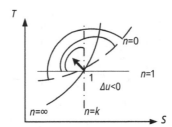

Fig. 4.24: Exercise 6(1): answer.

(2)

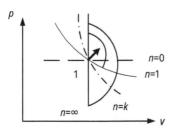

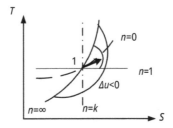

Fig. 4.25: Exercise 6(2): answer.

(3)

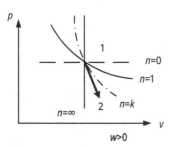

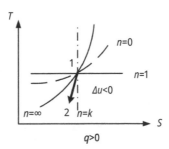

Fig. 4.26: Exercise 6(3): answer.

(4)

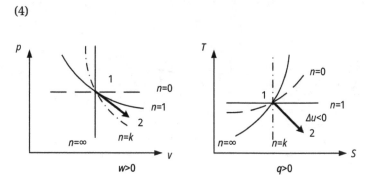

Fig. 4.27: Exercise 6(4): answer.

Chapter 5
Second law of thermodynamics

5.1 Review: the first law of thermodynamics

The first law of thermodynamics states that energy can neither be generated nor disappear out of nothing. It can only be transformed from one form to another or transferred from one object to another. In the process of transformation or transfer, the total amount of energy remains unchanged, that is, the law of conservation and transformation of energy.

The first law of thermodynamics describes a quantitative relationship between different types of energy. However, the direction, conditions, and limits of the process are not given.

5.2 Introduction to the second law of thermodynamics

1. What is the textual expression for the second law of thermodynamics?
2. How to judge whether a thermal process can occur, that is, what is the condition?
3. How to judge the conditions that any thermal process can be carried out, that is, what is the cost?
4. How to judge the limit of any thermal process, that is, the thermal efficiency?

5.3 The direction of spontaneous process and the expression of the second law of thermodynamics

5.3.1 Directionality and irreversibility of thermal process

As shown in Fig. 5.1, if the temperature of object A is higher than that of object B, then the heat transfer from B to A does not violate the first law. In other words, the heat transfer from a high-temperature object to a low-temperature object is spontaneous, and the heat transfer of a low-temperature object to a high-temperature object is non-spontaneous.

https://doi.org/10.1515/9783111329703-005

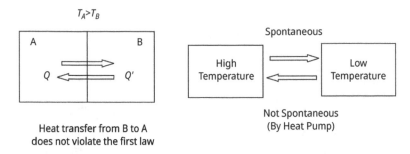

Heat transfer from B to A
does not violate the first law

Fig. 5.1: Heat transfer due to temperature difference.

Spontaneous process refers to the automatic process without any external action. Common spontaneous reactions in nature include:

1. Heat transfer due to temperature difference: Heat can only be transferred spontaneously (at no cost) from a high-temperature body to a low-temperature body.
2. Heat and work conversion: Work can be transformed into heat without cost.
   (1) Joule's heat work equivalent experiment: As shown in Fig. 5.2, the dropping of weight induces the water in the container to heat up, but does not allow the water to be cooled automatically and generate power to lift the weight.

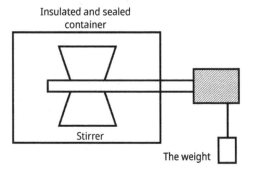

Fig. 5.2: Joule's heat work equivalent experiment.

   (2) Friction heat generation: Examples of friction heat generation include drilling wood for ignition and vehicle braking, as shown in Fig. 5.3.

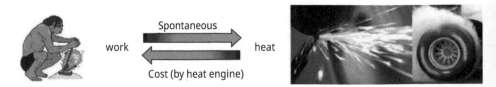

Fig. 5.3: Friction heat generation.

3. Free expansion: As shown in Fig. 5.4, matter can only flow spontaneously from the high-pressure space to the low-pressure space.

a.

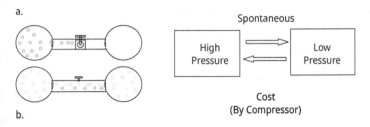

b.

Fig. 5.4: Schematic diagram for gas-free expansion.

4. Diffusive mixing: As shown in Fig. 5.5, the substance can only spontaneously diffuse from high-concentration liquid to low-concentration liquid.

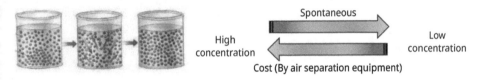

Fig. 5.5: Schematic diagram of diffusive mixing.

Let us do a summary:
1. Spontaneous process has directionality.
2. The reverse process of a spontaneous process is not impossible, but cannot be spontaneous. It can also be carried out if certain additional conditions are met or some changes are left to the surroundings.
3. Not all processes that do not violate the first law can be carried out.
4. The spontaneous process is irreversible, that is to say, all the spontaneous processes are irreversible.

Here is a question for you to think about: According to the first law of thermodynamics, energy is conserved and equivalent. How to explain the above thermodynamic phenomena? The answer is the second law of thermodynamics.

5.3.2 The expression of the second law of thermodynamics

Because of different kinds of thermal processes in nature, the second law of thermodynamics has many expressions, but they are equivalent to each other. Here are two famous expressions.

5.3.2.1 Way of heat transfer: Clausius statement in 1850

R. J. E. Clausius (1822–1888), a German physicist and mathematician, is one of the main founders of thermodynamics. As published in 1850, the basic concept of the second law of thermodynamics is clearly pointed out for the first time.

(1) It is impossible to construct a device that operates in a cycle and produces no effect other than transfer the heat from a lower temperature body to a higher temperature body, as shown in Fig. 5.6.

(2) Heat cannot be transferred spontaneously and without cost from a cold body to a hot body.

Compression refrigeration unit is a good example and heat transfers from low-temperature room to external high-temperature environment.

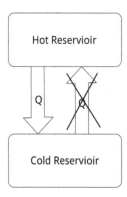

Fig. 5.6: Schematic diagram of the compression refrigeration unit.

That is to say, the spontaneous process of mechanical energy transfer to thermal energy is taken as the necessary cost to realize the heat transfer from the low-temperature object to high-temperature object.

5.3.2.2 Way of heat and work conversion: Kelvin-Planck statement in 1851

William Thomson, Lord Kelvin (1824–1907), a British mathematical physicist, is an inventor and one of the main founders of thermodynamics. The thermodynamic temperature scale was established in 1848, the second law of thermodynamics was proposed in 1851, and the Joule Thomson effect was discovered in 1852. Max Planck (1858–1947), a German physicist, is the founder of quantum mechanics, whose doctoral dissertation topic is around the second law of thermodynamics. He won the Nobel Prize in physics in 1918.

(1) It is impossible for any device that operates in a cycle to receive heat from a single reservoir and produce a net amount of work.

Ideal gas expansion process under the same temperature: $q = w$, with the pressure drops, the volume increases. As shown in Fig. 5.7, the state of the gas changes, or "has other influence." Therefore, it is not to say that heat cannot be converted into work completely, but can only be realized at the cost of other influence.

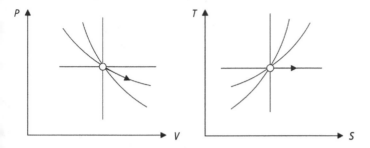

Fig. 5.7: Isothermal expansion of ideal gas.

(2) The second kind of perpetual motion machine is impossible to make success.
The principle of the second type of perpetual motion machine is shown in Fig. 5.8, and such perpetual motion machines do not violate the first law of thermodynamics. But it violates the second law of thermodynamics. Therefore, the second type of perpetual motion machine is impossible to manufacture successfully.

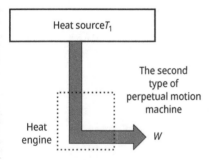

Fig. 5.8: The second kind of perpetual motion machine.

From these two statements, we can see the essence of the second law of thermodynamics as follows:
(1) Spontaneous processes are directional (directionality).
(2) The equivalence of statements is not an accident; it is a statement of common nature.
(3) If a process wants to reverse, it has to pay a cost (condition).

5.4 Carnot cycle and Carnot theorem

5.4.1 Thermodynamic cycle

Thermodynamic cycle: After a series of state changes, the working medium returns to the original state again (Fig. 5.9).

Reversible cycle: This is a cycle entirely consisting of reversible processes.

Irreversible cycle: Some or all of the processes in the cycle are irreversible [10].

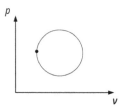

Fig. 5.9: Schematic diagram of a thermodynamic cycle.

The thermal cycle can be classified into forward cycle and reverse cycle.

(1) Forward cycle: As shown in Fig. 5.10, heat cannot be transferred spontaneously without cost from a cold body to a hot body. On the p–v and T–s graphs, the forward cycle is clockwise, and thus we have obtained this equation:

$$\oint dU = 0 \tag{5.1}$$

According to the first law of thermodynamics:

$$Q_1 - Q_2 = W_{net} \tag{5.2}$$

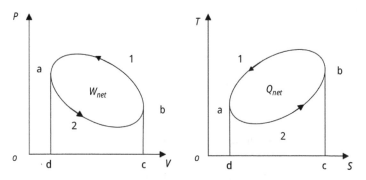

Fig. 5.10: Schematic diagram of the forward cycle.

Figure 5.11 shows how to calculate the thermal efficiency of the forward cycle. It refers to the ratio of the net work W_{net} done in the cycle to the heat Q_1 added to the working medium from the high-temperature heat source in the cycle:

$$\eta_t = \frac{W_{net}}{Q_1} = \frac{Q_1 - Q_2}{Q_1} = 1 - \frac{Q_2}{Q_1} \tag{5.3}$$

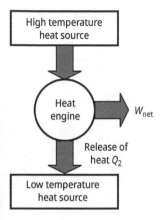

Fig. 5.11: Schematic diagram of thermal efficiency of the forward cycle.

The thermal efficiency of the cycle is used to evaluate the economy of the forward cycle. Clearly, $\eta_t < 1$.

(2) Reverse cycle: As shown in Fig. 5.12, heat is transferred from a cold reservoir to a hot reservoir in the cycle, such as a refrigeration cycle or a heat pump cycle.

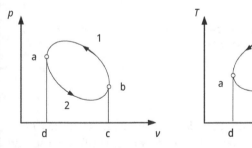

Fig. 5.12: Schematic diagram of a reverse cycle.

Through a reverse cycle:

$$\oint dU = 0 \tag{5.4}$$

According to the first law of thermodynamics:

$$Q_1 = Q_2 + W_{net} \tag{5.5}$$

Figure 5.13 illustrates the schematic diagram of the thermal efficiency of the reverse cycle. There are two coefficients: coefficient of refrigeration and coefficient of heat supply.

Coefficient of refrigeration is defined by

$$\varepsilon = \frac{Q_2}{W_{net}} = \frac{Q_2}{Q_1 - Q_2} \tag{5.6}$$

Coefficient of heat supply is defined by

$$\varepsilon' = \frac{Q_1}{W_{net}} = \frac{Q_1}{Q_1 - Q_2} \tag{5.7}$$

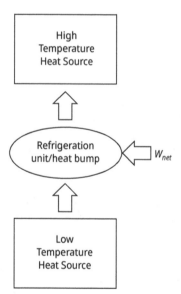

Fig. 5.13: Schematic diagram of the thermal efficiency of the reverse cycle.

5.4.2 Carnot cycle

The first law of thermodynamics verifies the first class of perpetual motion machines with $\eta_t > 100\%$ is impossible. The second law of thermodynamics verifies the second class of perpetual motion machines with $\eta_t = 100\%$ is impossible. A French engineer and scientist named N. L. S. Carnot (1796–1832) put forward an ideal heat engine working cycle in 1824. It consists of two reversible constant temperature processes and two reversible adiabatic processes. The concepts of "Carnot heat engine," "Carnot cycle," and "Carnot theorem" were put forward, which is the fundamentals for the second law of thermodynamics. Figure 5.14 shows an engine based on the principle of the Carnot cycle.

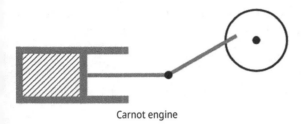

Carnot engine

Fig. 5.14: Schematic diagram of a Carnot engine.

Figure 5.15 illustrates the schematic diagram of the Carnot cycle, which is also called an ideal reversible heat engine cycle. The steps in the different stages are as follows:

a–b Constant temperature endothermic process:

$$q_1 = T_1(s_2 - s_1) \tag{5.8}$$

b–c Adiabatic expansion process, do work

c–d Constant temperature exothermic process:

$$q_2 = T_2(s_2 - s_1) \tag{5.9}$$

d–a Adiabatic compression process, consume work:

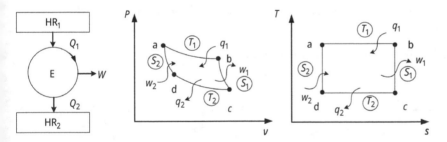

Fig. 5.15: Schematic diagram of the Carnot cycle.

The thermal efficiency of the Carnot cycle is defined by

$$\eta_c = 1 - \frac{q_2}{q_1} = 1 - \frac{T_2(s_2 - s_1)}{T_1(s_2 - s_1)} = 1 - \frac{T_2}{T_1} \tag{5.10}$$

Figure 5.16 shows the schematic diagram of the thermal efficiency of the Carnot cycle. The thermal efficiency of the Carnot cycle η_c only depends on the constant temperature heat sources T_1 and T_2, and not on the properties of the working medium.

As T_1 increases, η_c increases. As T_2 decreases, η_c increases. The greater the temperature difference, the higher the η_c.

When $T_1 \neq \infty$ or $T_2 \neq 0$ K, $\eta_c < 100\%$, that is, the second law of thermodynamics. When $T_1 = T_2$, $\eta_c = 0$, which means single source heat engine is impossible.

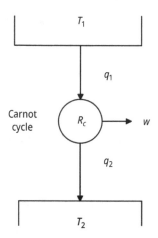

Fig. 5.16: The calculation for the thermal efficiency of the Carnot cycle.

5.4.3 Carnot theorem

Theorem 1: All reversible engines operating between the same hot and cold sources have the same thermal efficiency, independent of the properties of the working medium.

Figure 5.17 shows the proved procedures for Theorem 1.

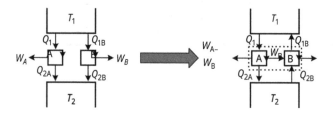

Fig. 5.17: Schematic diagram of Theorem 1.

Proof procedures: Two reversible heat engines (A and B) are set. The working fluids in A and B are different. The heat absorbed for the two reversible heat engine from the heat source is the same, both are Q_1:

$$W_A = Q_{1A} - Q_{2A} \tag{5.11}$$

$$W_B = Q_{1B} - Q_{2B} \tag{5.12}$$

$$\eta_A = \frac{W_A}{Q_{1A}} \tag{5.13}$$

$$\eta_B = \frac{W_B}{Q_{1B}} \tag{5.14}$$

Three possibilities are: (1) $\eta_A > \eta_B$, (2) $\eta_A < \eta_B$, (3) $\eta_A = \eta_B$. Let us prove cases (1) and (2) are false.

Proof by contradiction method: Suppose that $\eta_A > \eta_B$, because Q_1 is the same, so $W_A > W_B$, $Q_{2A} < Q_{2B}$. Reverse operation of B refrigerator is equivalent to taking out heat of low-temperature heat source $Q_{2B} - Q_{2A}$ and converting it into $W_A - W_B$.

For a single heat source engine, this violates the second law of thermodynamics. In the same way, it is impossible that $\eta_A < \eta_B$.

Theorem 2: The thermal efficiency of any irreversible engine operating between the same hot and cold sources is less than that of a reversible engine.

Figure 5.18 shows the energy consumption of reversible and irreversible engines.

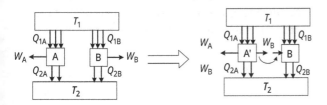

Fig. 5.18: Schematic diagram of Theorem 2.

Proof procedures: As shown in Fig. 5.18, let A be the irreversible engine and B be the reversible engine.

Proof by contradiction method, assume:

$$\eta_A = \eta_B \tag{5.15}$$

Let:

$$Q_{1A} = Q_{1B} \tag{5.16}$$

So,

$$W_A = W_B \tag{5.17}$$

$$Q_{2A} = Q_{2B} \tag{5.18}$$

Invert the reversible engine: The working medium cycle, and cold and heat sources are all restored to their original state, which means there are no trace outside and all the processes should be only reversible. But this contradicts the original assumption.

Assume:

$$\eta_A > \eta_B \tag{5.19}$$

Let:

$$Q_{1A} = Q_{1B} \tag{5.20}$$

$$W_A = Q_{1A} - Q_{2A} \tag{5.22}$$

$$W_B = Q_{1B} - Q_{2B} \tag{5.23}$$

$$W_A > W_B \tag{5.24}$$

$$Q_{2B} > Q_{2A} \tag{5.24}$$

Invert the reversible engine: $Q_{2B} - Q_{2A}$ means $W_A - W_B$. That equates to a single heat source engine, which violates the second law of thermodynamics.

Let us do a summary of the Carnot theorem: (1) Any reversible engine operating between two constant temperature sources at different temperatures T, $\eta_{tR} = \eta_{tc}$; (2) irreversible heat engine η_{t1R} < reversible heat engine working in the same heat source η_{tR}, $\eta_{t1R} < \eta_{tR} = \eta_{tc}$. All heat engines operating between given temperature limits: η_{tc} highest > Limit of heat engine. Figure 5.19 shows an example of the Carnot theorem.

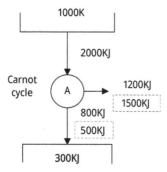

Fig. 5.19: An example of the Carnot cycle.

To check whether the heat engine can be realized, we can use the following equation:

$$\eta_{tc} = 1 - \frac{T_2}{T_1} = 1 - \frac{300}{1,000} = 70\%$$

$$\eta_t = \frac{w}{q_1} = \frac{1,200}{2,000} = 60\% \quad \text{Possible}$$

When $W = 1,500$ kJ

$$\eta_t = \frac{w}{q_1} = \frac{1,500}{2,000} = 70\% \quad \text{Impossible}$$

Comparison between actual cycle and Carnot cycle: the Carnot engine has only theoretical significance and the highest thermal efficiency, but it is hard to be realized in practice.

Internal combustion engine: $t_1 = 2{,}000$ °C, $t_2 = 300$ °C

$$\eta_{tc} = 74.7\%$$

Actually $\eta_t = 30 - 40\%$
Thermal power generation: $t_1 = 600$ °C, $t_2 = 25$ °C

$$\eta_{tc} = 65.9\%$$

Actually $\eta_t = 40\%$
By using heat recovery and combined cycle, η_t can reach 50%.

5.5 Entropy

5.5.1 Derivation of entropy

Carnot cycle:

$$\eta_c = 1 - \frac{|q_2|}{|q_1|} = 1 - \frac{T_2}{T_1} \tag{5.25}$$

Figure 5.20 shows the schematic diagram of the Carnot cycle. The derivation of entropy is as follows:

$$\frac{|q_2|}{|q_1|} = \frac{T_2}{T_1} \tag{5.26}$$

Change the absolute value to algebraic value:

$$\frac{q_1}{T_1} = -\frac{q_2}{T_2} \tag{5.27}$$

Then

$$\frac{q_1}{T_1} + \frac{q_2}{T_2} = 0 \tag{5.28}$$

or

$$\sum \frac{q}{T} = 0 \tag{5.29}$$

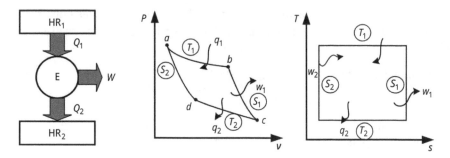

Fig. 5.20: Schematic diagram of the Carnot cycle.

Hence, the sum of the Carnot cycle $\frac{q}{T}$ is zero. Figure 5.21 shows a schematic diagram of a thermodynamic cycle.

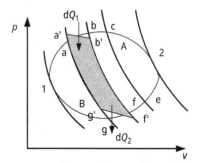

Fig. 5.21: Schematic diagram of a reversible cycle divided into microelement cycles.

The definition of entropy: Entropy change represents the direction and magnitude of heat exchange in a reversible process. It was first introduced by R. Clausius in the middle of the nineteenth century, where s was used to denote entropy since 1865 and professor Xianzhou Liu of Tsinghua University translated entropy into Chinese:

$$\Delta S = \int_1^2 ds = \int_1^2 \frac{\delta Q_{re}}{T} \tag{5.30}$$

$$dS = \frac{\delta Q_{re}}{T} \tag{5.31}$$

The specific entropy is defined by

$$ds = \frac{\delta q_{re}}{T} \tag{5.32}$$

Figure 5.22 shows two thermal cycles with different paths. By deriving the mathematic relations from eq. (5.33) to eq. (5.37), we come to the conclusion that entropy is a state

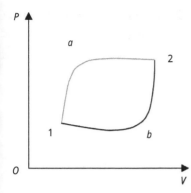

Fig. 5.22: Schematic diagram of two thermal cycles with different paths.

quantity. That is to say, entropy change has nothing to do with specific path, but it only depends on the initial and final states [11]:

$$\oint \frac{\delta Q}{T} = 0 \rightarrow \oint ds = 0 \tag{5.33}$$

$$\int_{1a2} \frac{\delta Q}{T} + \int_{2b1} \frac{\delta Q}{T} = 0 \tag{5.34}$$

$$\int_{2b1} \frac{\delta Q}{T} = -\int_{1b2} \frac{\delta Q}{T} \tag{5.35}$$

$$\int_{1a2} \frac{\delta Q}{T} = \int_{1b2} \frac{Q}{T} \tag{5.36}$$

$$\Delta S_{1a2} = \Delta S_{1b2} \tag{5.37}$$

5.5.2 Clausius inequality and entropy change of irreversible process

For reversible cycle, entropy is the defined by

$$dS = \frac{\delta Q_{re}}{T} \tag{5.38}$$

$$\oint ds = 0 \tag{5.39}$$

For irreversible cycle:

$$\oint \frac{\delta Q}{T} < 0 \tag{5.40}$$

It means that the working medium goes through any irreversible cycle, and the integral of $\frac{\delta Q}{T}$ over the entire cycle is negative. Hence, it is suitable for any irreversible cycle.

5.5.3 Expression of entropy in Clausius inequality and Clausius equality

5.5.3.1 Clausius inequality and Clausius equality
Clausius inequality and equality can be written as

$$\oint \frac{\delta Q}{T} \leq 0 \tag{5.41}$$

= Reversible cycle
< Irreversible cycle
> Impossible

This expression can be used as one of the mathematical expressions of the second law of thermodynamics. To judge whether a cycle can be carried out and whether it is reversible, this expression can be used as another expression for the second law of thermodynamics. The specific analysis method is given as follows based on the example shown in Fig. 5.23.

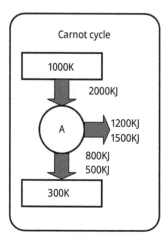

Fig. 5.23: An example to judge a thermodynamic process can occur or not.

To judge whether a heat engine can be realized or not, we first use the following equation:

$$\oint \frac{\delta Q}{T} = \frac{2,000}{1,000} - \frac{800}{300} = -0.667 \text{ kJ/K} < 0 \quad \text{possible}$$

If: $W = 1{,}500$ kJ

$$\oint \frac{\delta Q}{T} = \frac{2{,}000}{1{,}000} - \frac{500}{300} = 0.333 \text{ kJ/K} > 0 \quad \text{impossible}$$

It is worth mentioning that the positive and negative signs of heat need our attention.

5.5.3.2 Entropy changes of irreversible process

Figure 5.24 shows the schematic diagram of entropy change for reversible process (solid curve) and irreversible process (dash curve).

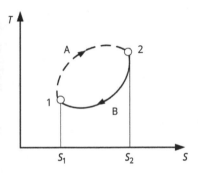

Fig. 5.24: Schematic diagram of entropy change for irreversible process.

For any irreversible cycle, according to the Clausius inequality

$$\oint \frac{\delta Q}{T} < 0 \tag{5.42}$$

$$\int_{1A2} \frac{\delta Q}{T} + \int_{2B1} \frac{\delta Q}{T} < 0 \tag{5.43}$$

$$\int_{2B1} \frac{\delta Q}{T} = - \int_{1B2} \frac{\delta Q}{T} \tag{5.44}$$

$$\int_{1A2} \frac{\delta Q}{T} < \int_{1B2} \frac{\delta Q}{T} = \Delta S_{12} \tag{5.45}$$

thus

$$\Delta S_{12} = S_1 - S_2 \geq \int_{12} \frac{\delta Q}{T} \tag{5.46}$$

= Reversible
> Irreversible

5.5.3.3 The relationship between ΔS and heat flow

$$\Delta S_{12} = S_1 - S_2 \geq \int_{12} \frac{\delta Q}{T} \tag{5.47}$$

$= $ Reversible
$> $ Irreversible
$< $ Impossible

(1) For a thermodynamic cycle:

$$\Delta S_{12} = 0 \tag{5.48}$$

(2) For a thermodynamic process:

$$0 \geq \oint \frac{\delta Q}{T} \tag{5.49}$$

Besides heat transfer, there are other factors that affect entropy:

$$\Delta S \geq \oint \frac{\delta Q}{T} \tag{5.50}$$

Irreversible adiabatic process: $\delta Q = 0$
Irreversible factors can cause entropy changes: $dS > 0$
Hence, the irreversible process is always companied by entropy increasing.

5.5.3.4 Entropy flow and entropy production

For any infinitesimal process:
(1) Entropy flow, which is the change in entropy caused by heat exchange between a working medium and a heat source

$$dS_f = \frac{\delta Q}{T} \tag{5.51}$$

(2) Entropy generation, which is caused by purely irreversible factors

$$dS_g > 0 \tag{5.52}$$

$$dS = dS_f + dS_g \tag{5.53}$$

$$\Delta S = \Delta S_f + \Delta S_g \tag{5.54}$$

Now we can come to the conclusion that entropy generation is a measure of the irreversibility of a process.

(3) The relationship among entropy change, entropy flow, and entropy generation. Tab. 5.1 shows the changes of them in different processes:

$$dS = dS_f + dS_g \tag{5.55}$$

$$\Delta S = \Delta S_f + \Delta S_g \tag{5.56}$$

Tab. 5.1: Entropy change, entropy flow, and entropy generation for different processes.

Any irreversible process	$\Delta S \geq 0$	$\Delta S_f \geq 0$	$\Delta S_g > 0$
Reversible process	$\Delta S = \Delta S_f \geq 0$	$\Delta S_f \geq 0$	$\Delta S_g = 0$
Irreversible adiabatic process	$\Delta S > 0$	$\Delta S_f = 0$	$\Delta S_g > 0$
Reversible adiabatic process	$\Delta S = 0$	$\Delta S_f = 0$	$\Delta S_g = 0$

Next we are going to introduce the principle of entropy increases and the loss of work capacity of an isolated system. The principle of entropy increases of an isolated system can be expressed as follows:

For an isolated system:

$$dS_g = 0 \tag{5.57}$$

If there is no mass exchange, no heat exchange, and no work exchange for the system:

$$dS_{ios} = dS_g \geq 0 \tag{5.58}$$

$$= \text{reversible process}$$
$$> \text{irreversible process}$$

This is another expression for the second law of thermodynamics. We can come to the conclusion that the entropy of an isolated system can only increase, or not change, but never decrease. This law is called the principle of entropy increase of an isolated system.

The loss of work capacity is another important definition here. Work capacity refers to the maximum useful work that the system can make when it reaches the thermal equilibrium with the environment. For any system undergoes an irreversible process, the loss of work capacity results in the increase in the entropy of its isolated system. The relationship between the loss of work capacity and the entropy increase of an isolated system is expressed as

$$I = T_0 \Delta S_{ios} \tag{5.59}$$

When the ambient temperature T_0 is determined, the loss of work capacity I is directly proportional to the entropy increase of the isolated system ΔS_{ios}. The entropy increase of an isolated system is a measure of work capacity loss. The above formula is applicable to the calculation of the loss of working capacity caused by any irreversible factor.

5.6 Summary

In this chapter, we have learned the following points:
1. Essence and expression of the second law of thermodynamics
2. Forward and reverse cycle evaluation indexes
3. Carnot cycle and Carnot theorem
4. The derivation and calculation of entropy and Clausius inequality
5. The principle of entropy increase of an isolated system
6. The application of the second law of thermodynamics to the judgment of directivity and reversibility

Exercises

Please choose true or false for the following questions:
1. The process of entropy increase is irreversible.
    T: True   F: False
2. The entropy change ΔS of an irreversible process cannot be calculated.
    T: True   F: False
3. The process of nature is in the direction of entropy increase, so the process of entropy decrease is impossible to realize.
    T: True   F: False
4. The entropy of the heated working medium must increase, and the entropy of the exothermic working medium must decrease.
    T: True   F: False
5. If the working medium reaches the same final state from a certain initial state through reversible and irreversible paths, the ΔS of irreversible path must be greater than that of reversible path.
    T: True   F: False
6. $\Delta S > 0$ is valid after working medium undergoes an irreversible cycle.
    T: True   F: False
7. The thermal efficiency increases with the increased network output of the cycle. Is the thermal efficiency for all reversible cycles equal? Is the thermal efficiency of irreversible cycle less than that of reversible cycle? Why?
8. What are the scope of application for the two following formulas for the calculation of thermal efficiency for a thermodynamic cycle $\eta_t = \frac{q_2}{q_1}$ and $\eta_t = \frac{T_2}{T_1}$?

9. A reversible heat engine works between a high-temperature heat source with a temperature of 150 °C and a low-temperature heat source with a temperature of 20 °C. Try to find out:
   (1) What is the thermal efficiency of the heat engine?
   (2) When the output work of the heat engine is 3.2 kJ, what is the heat absorbed from the high-temperature heat source and the heat released to the low-temperature heat source?

10. An inventor claims to have designed a heat engine that circulates between 500 and 300 K heat sources. The heat engine can make 450 J network for every 1,000 J heat absorbed from high-temperature heat sources. Is his design reasonable?

Answers

1. F
2. F
3. F
4. F
5. F
6. F
7. The thermal efficiency is the ratio of the output network to the amount of heat absorbed. Therefore, under the same condition of heat absorption, the greater the net output of the cycle output, the higher the thermal efficiency. Not all reversible cycles have the same thermal efficiency. For the same initial and final states, the thermal efficiency of irreversible cycle must be less than that of the reversible cycle.

8. The former formula is applicable to all kinds of reversible and irreversible cycles, while the latter is only applicable to the reversible Carnot cycle.

9. (1) $\eta = \dfrac{T_1 - T_2}{T_1} = \dfrac{150 - 20}{423} = 30.7\%$

   (2) $Q_1 = \dfrac{W}{\eta_1} = \dfrac{3.2}{30.7\%} = 10.42\,\text{kJ}$

   $Q_2 = Q_1 - W = 10.42 - 3.2 = 7.22\,\text{kJ}$

10. Approach 1

   $\eta_{tc} = 1 - \dfrac{T_2}{T_1} = 1 - \dfrac{300}{500} = 40\%$

   $\eta_t = \dfrac{W}{Q_1} = \dfrac{450}{1,000} = 45\% > \eta_{tc}$ unreasonable

Approach 2

$$Q_2 = Q_1 - W = 1{,}000 - 450 = 550 \text{ kJ}$$

$$\frac{Q_1}{T_1} - \frac{Q_2}{T_2} = \frac{1{,}000}{500} - \frac{550}{300} = 0.1667 \text{ kJ/K} > 0 \quad \text{unreasonable}$$

Chapter 6
Water vapor and wet air

6.1 Review: gaseous working medium

(1) Gas: Away from liquid, generally can be treated as an ideal gas, such as air.
(2) Vapor: Close to liquid, generally not as an ideal gas, such as water vapor, ammonia vapor, and fluorine vapor.

6.2 Review: water vapor

Water vapor cannot be treated as an ideal gas but can only be treated as actual gas, and thus we need to use the chart to determine its state.

When steam engine was invented in the eighteenth century, water vapor was the only working medium. Until the invention of the internal combustion engine, there was no gas working medium. Nowadays, it is still a working medium for thermal power generation, nuclear power, heating, and chemical engineering. There are many advantages, such as cheap, easily available, non-toxic, good swelling performance, and good heat transfer performance.

This chapter analyzes its generation, status determination, and basic thermal processes.

6.3 Water vapor generation process

6.3.1 Basic concepts

Vapor is produced by the vaporization of liquids. The two forms of liquid vaporization are evaporation and boiling.

(1) Evaporation:
Vaporization phenomenon on the surface of a liquid occurs under any temperature, as shown in Fig. 6.1. The higher the temperature is, the more intense the evaporation is.

(2) Boiling:
Under a given pressure, boiling occurs at a certain temperature, simultaneously inside and on the surface of the liquid. Boiling accompanies by a large amount of bubbles, as shown in Fig. 6.2.

https://doi.org/10.1515/9783111329703-006

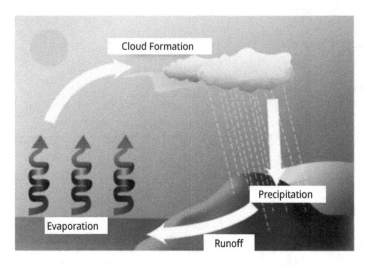

Fig. 6.1: Diagram of vaporization phenomenon.

Fig. 6.2: Diagram of a boiling phenomenon.

Saturation will occur in evaporation and boiling. However, not every evaporation will reach saturation. The following two situations will be introduced according to the different containers in which the evaporation is located.

(1) Vaporization in an open container: Continue until all liquids become vapor, as shown in Fig. 6.3.

(2) Vaporization in a closed container: When the number of molecules escaping from the liquid surface is equal to the number of molecules returning to the liquid surface, the gas and liquid phases reach dynamic equilibrium, and this state is saturation, as shown in Fig. 6.4.

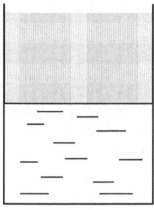

Untill all liquids become
vapor

Fig. 6.3: Diagram of vaporization in an open container.

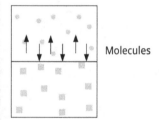

Molecules

Fig. 6.4: Diagram of vaporization in a closed container.

Liquids and vapors in a saturated state are called saturated liquids and saturated vapors, respectively. The pressure and temperature of saturated vapor are called saturation pressure (represented by P_s) and saturation temperature (represented by T_s). The two correspond to each other. The higher the saturation pressure, the higher the saturation temperature, as shown in Fig. 6.5. Examples are as follows:

Standard atmospheric pressure: p_s = 1.01325 bar, t_s = 100 °C
Tibetan atmospheric pressure: p_s = 0.8 bar, t_s = 93 °C
High pressure: p_s = 1.6 bar, t_s = 113.3 °C

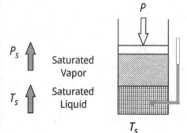

Fig. 6.5: Diagram of parameter of saturation.

6.3.2 Isobaric heating process of water (five states)

Water vapor used in engineering is usually produced by heating water in a boiler at a constant pressure. Figure 6.6 is used to illustrate the generation process of water vapor: suppose a cylindrical container contains 1 kg of water at temperature t, there is a movable piston on the water surface. A certain pressure p can be exerted on the piston and the water is heated at the bottom of the container. The production process of water vapor can generally be divided into the following three stages: water constant pressure heating (Fig. 6.6a,b), saturated water constant pressure vaporization (Fig. 6.6b,d), and dry steam constant pressure superheat (Fig. 6.6d,e).

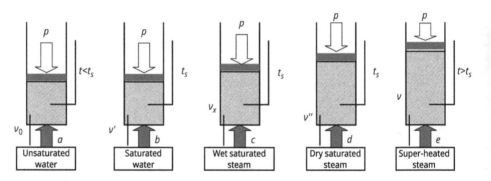

Fig. 6.6: Diagram of heating water in a boiler at a constant pressure.

(1) Evaporation latent heat:

$$r = h'' - h' \tag{6.1}$$

(2) Degree of super-heat:

$$t - t_s \tag{6.2}$$

The constant pressure formation process of water vapor can be illustrated on the p–v diagram and the T–s diagram, as shown in Fig. 6.7. On the p–v diagram, it is a horizontal line; a–b, b–d, and d–e are the processes of constant pressure preheating, constant pressure vaporization, and constant pressure overheating, respectively. The a, b, c, d, and e, respectively, represent the five states corresponding to Fig. 6.6. On the T–s diagram, the constant pressure formation process line A-B-C-D-E of water vapor is divided into three sections. A–B is the constant pressure preheating process, the temperature increases during the process, the entropy increases, and the process line slopes upward to the right; B–D is the constant pressure vaporization process. During the process, the pressure and temperature remain unchanged, and the entropy increase, which is a horizontal line on the T–s diagram; D–E is the constant pressure superheating process, the temperature increases, the entropy increases, and the process line slopes upward to the right. The total heat absorbed by the working fluid is represented by the area under the A-B-C-D-E process line on the T–s diagram.

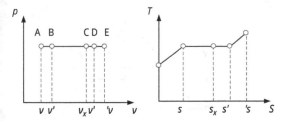

Fig. 6.7: Isobaric heating process of water in p–v and T–s diagram.

As shown in Fig. 6.8, when the pressure increases, the distance between saturated water and dry saturated steam decreases. When the pressure increases to a certain critical value, the two points coincide, and there is no difference between saturated water and dry saturated steam. This special state is called a critical state, as point C in Fig. 6.8.

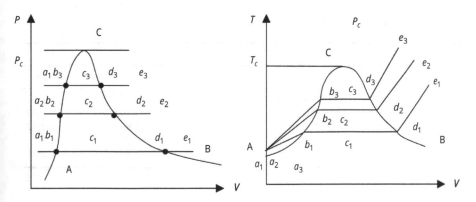

Fig. 6.8: The process of water vapor generation at different pressures.

When water is heated to a critical temperature at a critical pressure or above, as shown in Fig. 6.9, there is no vapor–liquid boundary line and vaporization process where vapor–liquid phases coexist, and directly become superheated steam if keep heating. The critical parameters of water are: p_{cr} = 22.064 MPa, t_{cr} = 373.99 °C, v_{cr} = 0.003104 m³/h.

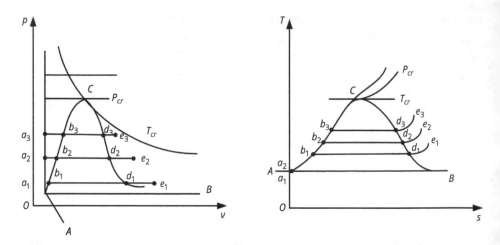

Fig. 6.9: Diagram of critical parameters of water.

The saturation parameters on the isobar line are shown in Tab. 6.1.

Tab. 6.1: Saturation parameters on the isobaric line.

p	t_s	v'	v''	s'	s''
(bar)	(°C)	(m³/kg)	(m³/kg)	kJ/(kg · K)	kJ/(kg · K)
0.006112	0.01	0.00100022	206.175	0.0	9.1562
1.0	99.63	0.0010432	1.6943	1.3028	7.3589
5.0	151.87	0.0010925	0.37486	1.8610	6.8216
50.0	263.98	0.0012862	0.039439	2.9201	5.9724
220.64	373.99	0.003106	0.003106	4.4092	4.4092

6.4 State parameters of water vapor

The basic principles for determining the parameters of water and water vapor states are:

Unsaturated water and superheated steam: Determine any two independent parameters, such as p and T.

Saturated water and dry saturated steam: Determine p or T.

Wet saturated steam: Other parameters are related to the two-phase ratio except p or T.

The diagram of the critical parameters of water is shown in Fig. 6.10.

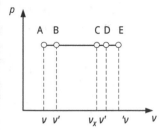

Fig. 6.10: Diagram of critical parameters of water.

6.4.1 Determination of the state parameters of the wet saturated steam zone

Since the temperature and pressure of wet steam are two interdependent parameters, so giving the temperature and pressure of the wet steam does not determine the state of the wet steam. Since wet steam is composed of dry saturated steam and saturated water at the same pressure and temperature, it is composed of different mass ratios, so it is necessary to specifically determine the state of wet steam. In addition to stating its pressure or temperature, the mass ratio of dry saturated steam to saturated water must also be indicated. Thus, it is necessary to introduce the dryness for calculation. The mass fraction of dry saturated steam contained in wet steam is called the dryness of wet steam, which is represented by x. The formula is as follows:

$$x = \frac{\text{Quality of dry saturated steam}}{\text{Quality of wet saturated steam}} = \frac{m_v}{m_v + m_f} \qquad (6.3)$$

where m_v denotes dry saturated steam and m_f denotes saturated water.
The state of matter represented by different dryness x is as follows:

$$x = 0: \text{ saturated water}$$

$$x = 1: \text{ dry saturated steam}$$

$$0 \le x \le 1 \text{ wet vapor}$$

It is worth mentioning that in unsaturated water and superheated steam areas, x is meaningless.

If there is 1.0 kg wet vapor and the dryness is x, then there is x kg dry saturated steam and $(1-x)$ kg saturated water. Therefore, the relevant parameters of 1 kg of wet steam are equal to the sum of the corresponding parameters of x kg of dry steam and the corresponding parameters of $(1-x)$ kg of saturated water:

$$h = xh'' + (1-x)h' \qquad (6.4)$$

$$v = xv'' + (1-x)v' \qquad (6.5)$$

$$s = xs'' + (1-x)s' \qquad (6.6)$$

The unit of h is kJ/kg, so the unit of p should be kPa. Using the above relationship, when the pressure (temperature) of the wet steam and a certain parameter y are known, its dryness can be determined:

$$y = xy'' + (1-x)y' \qquad (6.7)$$

$$x = \frac{y - y'}{y'' - y'} \qquad (6.8)$$

6.4.2 Water and water vapor meters

In general, the nature of water vapor is greatly different from the ideal gas. For the convenience of engineering calculations, list the state parameters of unsaturated water, saturated water, dry saturated steam, and superheated steam under different T and p into a table or draw a line graph [12]. Two categories: (1) saturated water and saturated steam (Tabs. 6.2 and 6.3), and (2) unsaturated water and superheated steam (Tab. 6.4).

6.4.2.1 Saturated water and saturated steam

Tab. 6.2: Thermodynamic properties of saturated water and saturated water vapor (sorted by temperature).

Temperature	Pressure	Specific volume		Specific enthalpy		Latent heat of vaporization	Specific entropy	
		Liquid	Vapor	Liquid	Vapor		Liquid	Vapor
t	p	v'	v''	h'	h''	r	s'	s''
°C	MPa	$\frac{m^3}{kg}$	$\frac{m^3}{kg}$	$\frac{kJ}{kg}$	$\frac{kJ}{kg}$	$\frac{kJ}{kg}$	$\frac{kJ}{kg \cdot K}$	$\frac{kJ}{kg \cdot K}$
0.00	0.0006112	0.00100022	206.154	−0.05	2500.51	2500.6	−0.0002	9.1544
0.01	0.0006117	0.00100021	206.012	0.00	2500.53	2500.5	0.0000	9.1541
1	0.0006571	0.00100018	192.464	4.18	2502.35	2498.2	0.0153	9.1278
2	0.0007059	0.00100013	179.787	8.39	2504.19	2495.8	0.0306	9.1014
3	0.0007580	0.00100009	168.041	12.61	2506.03	2493.4	0.0459	9.0752

Tab. 6.3: Thermodynamic properties of saturated water and saturated water vapor (sorted by pressure).

Pressure	Temperature	Specific volume		Specific enthalpy		Latent heat of vaporization	Specific entropy	
		Liquid	Vapor	Liquid	Vapor		Liquid	Vapor
p	t	v'	v''	h'	h''	r	s'	s''
MPa	°C	$\dfrac{m^3}{kg}$	$\dfrac{m^3}{kg}$	$\dfrac{kJ}{kg}$	$\dfrac{kJ}{kg}$	$\dfrac{kJ}{kg}$	$\dfrac{kJ}{kg \cdot K}$	$\dfrac{kJ}{kg \cdot K}$
0.0010	6.9491	0.0010001	129.185	29.21	2513.29	2484.1	0.1056	8.9735
0.0020	17.5403	0.0010014	67.008	73.58	2532.71	2459.1	0.2611	8.7220
0.0030	24.1142	0.0010028	45.666	101.07	2544.68	2443.6	0.3546	8.5758
0.0040	28.9533	0.0010041	34.796	121.30	2553.45	2432.2	0.4221	8.4725
0.0050	32.8793	0.0010053	28.191	137.72	2560.55	2422.8	0.4761	8.3930

6.4.2.2 Unsaturated water and superheated steam

Tab. 6.4: Thermodynamic properties of unsaturated water and superheated steam.

p	0.001 MPa (t_s = 6.949 °C)			0.005 MPa (t_s = 32.879 °C)		
	v'	h'	s'	v'	h'	s'
	0.001001	29.21	0.1056	0.0010053	137.72	0.4761
	m^3/kg	kJ/kg	kJ/(kg · K)	m^3/kg	kJ/kg	kJ/(kg · K)
	v''	h''	s''	v''	h''	s''
	129.185	2513.3	8.9735	28.191	2560.6	8.3930
	m^3/kg	kJ/kg	kJ/(kg · K)	m^3/kg	kJ/kg	kJ/(kg · K)
t	v	h	s	v	h	s
°C	$\dfrac{m^3}{kg}$	$\dfrac{kJ}{kg}$	$\dfrac{kJ}{kg \cdot K}$	$\dfrac{m^3}{kg}$	$\dfrac{kJ}{kg}$	$\dfrac{kJ}{kg \cdot K}$
0	0.001002	−0.05	−0.0002	0.0010002	−0.05	−0.0002
10	130.598	2519.0	8.993 8	0.0010003	42.01	0.1510
20	135.226	2537.7	9.058 8	0.0010018	83.87	0.2963
40	144.475	2575.2	9.182 3	28.854	2 574.0	8.4366
60	153.717	2612.7	9.298 4	30.712	2611.8	8.5537
80	162.956	2650.3	9.408 0	32.566	2 649.7	8.6639
100	172.192	2688.0	9.512 0	34.418	2 687.5	8.7682

Zero-point rules in the table: International regulations, for the steam three-phase point (solid, liquid, and steam), the liquid-phase water thermodynamics u and entropy s are set to be zero.

That is to say, at the three-phase point, the state parameters of liquid water are:

$$p = 611.7 \text{ Pa}$$

$$v = 0.00100021 \text{ m}^3/\text{kg}$$

$$T = 273.16 \, K$$

$$u = 0 \, \text{kJ/kg}$$

$$s = 0 \, \text{kJ}/(\text{kg} \cdot \text{K})$$

$$h = u + p \, v = 0.0061 \, \text{kJ/kg} \approx 0 \, \text{kJ/kg}$$

Example:
(1) Need to first determine the state belongs to which of the five states: p = 1 MPa, determine the state and h at t = 100 °C, 200 °C?

$$t_s(p) = 179.916 \text{ °C}$$

$$t = 100 \text{ °C} < t_s \text{ unsaturated water}$$

$$h = 419.7 \, \text{kJ/kg}$$

$$t = 200 \text{ °C} > t_s \text{ overheated steam}$$

$$h = 2{,}827.3 \, \text{kJ/kg}$$

(2) Fill a rigid container with p = 0.1 MPa, t = 20 °C water, due to Sun's exposure, its temperature rises to 40 °C. Hence, please determine the pressure on containers:
Isovolumic process: p = 0.1 MPa, t = 20 °C
v = 0.0010018 m³/kg (unsaturated water)
When t = 40 °C

$$p = 14.0 \text{ MPa}$$

The reservoir is dangerous and cannot be filled full.
(3) Known t = 83 °C, p = 0.01MPa, please determine the status. h = ?
$p = 0.01 \, MPa, \quad t_s = 45.8 \text{ °C (overheating state)}$
Interpolation:

$$\frac{h - h_{80}}{h_{100} - h_{80}} = \frac{83 - 80}{100 - 80} \tag{6.9}$$

$$h_{80} = 2{,}648.9 \, \text{kJ/kg}$$

$$h_{100} = 2{,}686.9 \, \text{kJ/kg}$$

$$h_{83} = 2{,}654.6 \, \text{kJ/kg}$$

6.5 Basic thermal process of water vapor

6.5.1 Thermal process of water vapor

Thermal process will revolve around the four physical quantities of p, s, T, and v.

The task that needs to be carried out is as follows: (1) determining the parameters of the initial and final states; (2) showing them in p–v, T–s, h–s figures; and (3) calculating work and heat in the process.

In thermal calculations, the most encountered situations are:

The isobaric process of water vapor (such as the generation of water vapor in a boiler and the condensation of water vapor in a condenser)

The isentropic process (the expansion of water vapor in a steam engine or a steam turbine)

The first law and the second law of thermodynamics are both valid [11]:

$$\delta q = du + \delta w \tag{6.10}$$

$$\delta q = dh + \delta w_t \tag{6.11}$$

Quasi-static:

$$\delta w = pdv \tag{6.12}$$

Reversible:

$$\delta w_t = -vdp \tag{6.13}$$

The characteristic properties and expressions of the ideal gas cannot be used:

$$pv = RT \tag{6.14}$$

$$u = f(T) \tag{6.15}$$

$$h = f(T) \tag{6.16}$$

$$c_p - c_v = R \tag{6.17}$$

$$c_p = \frac{k}{k-1} R \tag{6.18}$$

$$c_v = \frac{1}{k-1} R \tag{6.19}$$

$$\Delta s = c_p \ln \frac{T_2}{T_1} - R \ln \frac{P_2}{P_1} \tag{6.20}$$

A schematic diagram of the steam power cycle is shown in Fig. 6.11, and its thermal process formula is as follows:

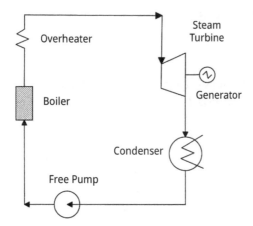

Fig. 6.11: Schematic diagram of steam power cycle.

Isobaric process:

$$q = \Delta h + w_t = \Delta h \tag{6.21}$$

Isentropic process:

$$w_t = -\Delta h \tag{6.22}$$

6.5.2 Isobaric process of water vapor

6.5.2.1 Boilers and condensers

$$q = \Delta h \tag{6.23}$$

$$w_t = 0 \tag{6.24}$$

Example: In the boiler, water is heated from 20 °C, 3 MPa, to 450 °C in isobaric process. Please show its process in the graph. Δh = ?

According to the parameters of the initial state and final state, the p–v diagram (Fig. 6.12), T–s diagram, and h–s diagram (Fig. 6.13) can be determined:

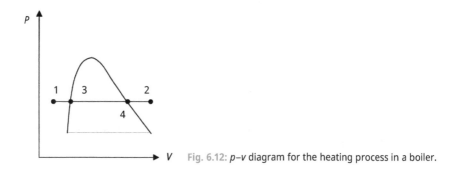

Fig. 6.12: p–v diagram for the heating process in a boiler.

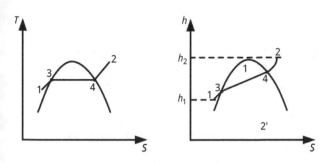

Fig. 6.13: *T–s* and *h–s* diagram for the heating process in a boiler.

$$t_s(3\text{ MPa}) = 232.893\text{ °C}$$

$$h_1 = 86.68\text{ kJ/kg}$$

$$h_2 = 3,343.0\text{ kJ/kg}$$

$$q = h_2 - h_1 = 3,256.32\text{ kJ/kg}$$

6.5.2.2 Turbines and feed pumps ($q = 0$)

The analysis process is the same as the analysis process of the above boiler and condenser: first, draw the *p–v* diagram (Fig. 6.14), *T–s* diagram (Fig. 6.15), and *h–s* diagram (Fig. 6.16):

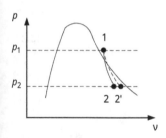

Fig. 6.14: *p–v* diagram for the turbines and feed pumps.

$$w_t = -\Delta h = h_2 - h_1 \tag{6.25}$$

Reversible process: $1 \rightarrow 2$
Irreversible process: $1 \rightarrow 2'$

Reversible process:

$$w_t = h_1 - h_2 \tag{6.26}$$

Irreversible process:

$$w_t = h_1 - h_{2'} \tag{6.27}$$

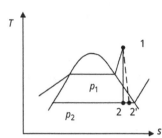

Fig. 6.15: *T–s* diagram for the turbines and feed pumps.

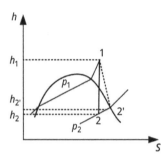

Fig. 6.16: *h–s* diagram for the turbines and feed pumps.

Reversible process:

$$w_t = h_1 - h_2 \tag{6.26}$$

Irreversible process:

$$w_t = h_1 - h_{2'} \tag{6.27}$$

Turbine efficiency:

$$\eta_{oi} = \frac{h_1 - h_{2'}}{h_1 - h_2} \tag{6.28}$$

6.6 The nature of wet air

6.6.1 Basic definitions

6.6.1.1 Wet air and dry air
Wet air: Air containing water vapor
Dry air: Air completely free of water vapor
    Water vapor can be regarded as an ideal gas since it is at low partial pressure. In general, humid air can be regarded as an ideal mixed gas.

According to Dalton's law, total pressure of wet air p is equal to the sum of partial pressure of water vapor p_v and partial pressure of dry air p_a:

$$p = p_v + p_a \qquad (6.29)$$

6.6.1.2 Unsaturated wet air and saturated wet air

Unsaturated wet air: The water vapor in the wet air is unsaturated and overheated, and the wet air absorbs the water, $p_v < p_s(T)$ as point 1 in Fig. 6.17.

Saturated wet air: Water vapor in wet air is saturated and can no longer absorb water, $p_v = p_s(T)$, as point 3.

In constant temperature hygroscopic processes 1 to 3, unsaturated wet air transforms into saturated wet air.

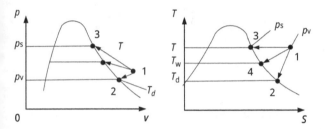

Fig. 6.17: The state of water vapor in wet air.

6.6.1.3 Humidity, relative humidity, absolute humidity, moisture content

Humidity: Content of water vapor in wet air

Absolute humidity: Density of water vapor in wet air

$$\rho_v = \frac{m_v}{V} = \frac{P_v}{R_{g,v}T} \qquad (6.30)$$

$$\rho_s = \frac{m_s}{V} = \frac{P_s}{R_{g,v}T} \qquad (6.31)$$

Relative humidity: The ratio of the absolute humidity of water vapor in wet air to the maximum absolute humidity at the same temperature

$$\varphi = \frac{\rho_v}{\rho_s} = \frac{P_v}{P_s} \qquad (6.32)$$

Moisture content (d): Water vapor mass in dry air per unit mass

$$d = \frac{m_v}{m_s} = \frac{\rho_v}{\rho_s} = \frac{P_v/R_{g,v}T}{P_s/R_{g,v}T} = 0.622\frac{P_v}{P-P_v} = 0.622\frac{\varphi P_s}{P-\varphi P_s} \qquad (6.33)$$

6.6.2 Dew point (T_d)

Dew: Cooling the wet air to the point in isobaric process if keep cooling, part of the water vapor will condense into water (Fig. 6.18, processes 1 and 2)

Dew point temperature (T_d): Corresponding saturation temperature of water vapor pressure in wet air P_v.

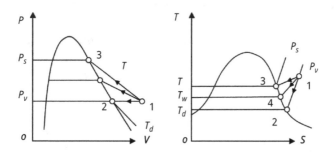

Fig. 6.18: The state of water vapor in wet air.

Examples:

If the temperature of the air is 20 °C, pressure is 0.1 MPa, and relative humidity is 60%, what temperature should it be when the dew appears?

Solution: From the table of saturated water and saturated steam (Appendix Table A1), the saturation pressure of water vapor at 20 °C is:

$$P_s = 0.023385 \times 10^5 \text{ Pa}$$

According to eq. (6.32), the partial pressure of water vapor in air is

$$P_v = \varphi P_s = 0.60 \times 0.023385 \times 10^5 \text{ Pa} = 0.01403 \times 10^5 \text{ Pa}$$

Checking the appendix, Table A2 shows that the saturation temperature corresponding to this pressure is 11.2 °C, which is the dew point. When the temperature drops to 11.2 °C, dew will appear.

6.7 Summary

(1) Grasp the process of isobaric heating of water vapor, and can show the state of water vapor during the process on p–v diagram and T–s diagram.
(2) Determine the state parameters of water vapor by using the water and water vapor table, H–s diagram of water vapor.
(3) Concept: Wet air, dry air, dew point, humidity, relative humidity, absolute humidity, and moisture content.

Exercises

1. Use the water vapor meter to determine the states at the following points, and determine the enthalpy, entropy, or dryness and specific volume of each state:
   (1) $p = 20$MPa, $t = 300$ °C
   (2) $p = 9$ MPa, $v = 0.017$ m³/kg
2. The boiler produces 20 t/h of steam, its pressure is 4.55 MPa, and the temperature is 480 °C. The water entering the boiler has a pressure of 4.5 MPa and a temperature of 30 °C. If the boiler efficiency rate is 0.8 and the calorific value of coal is 30,000 kJ/kg, how much coal is needed in 1 h (boiler efficiency is the ratio of the total heat absorbed by the steam to the total heat generated by the fuel)?
3. In the steam turbine, the initial steam parameters are: $p_1 = 3$ MPa and $t_1 = 350$ °C. If the reversible adiabatic expansion reaches $p_2 = 0.006$ MPa and the steam flow is 3.4 kg/s, please find the ideal power of the steam turbine.
4. Turbine inlet parameters: $p_1 = 4.0$ MPa, $t_1 = 450$ °C, outlet pressure $p_2 = 5$ kPa, steam dryness $x_2 = 0.9$, calculate the relative internal efficiency of the steam turbine.
5. The exhausted steam of the steam turbine enters the condenser under the state of vacuum degree of 0.094 MPa and $x = 0.90$. Constant pressure cooling condenses into saturated water. Try to calculate the multiple of the volume reduction when the exhausted steam condenses into water, and find the heat released by 1 kg of exhausted steam in the condenser. The atmospheric pressure is known to be 0.1 MPa.
6. A rigid closed container is filled with 0.1 MPa, 20 kg of water at 20 °C. For example, due to accidental heating, the temperature rises to 40 °C:
   (1) Find the heat added to generate this temperature rise.
   (2) Coping with this contingency, how much pressure should the vessel be safe?

Answers

1. (1) Unsaturated water and (2) wet saturated steam. Other parameters are shown in Tab. 6.4.

Tab. 6.4: Parameters for Exercise 1.

p (MPa)	20	9
t (°C)	300	303.385
x	0	0.8172
v	0.001305	0.017
h (kJ/kg)	1333.4	2489.9
s (kJ/kg · K)	3.2072	5.2399

2. First check the steam table or diagram to determine the boiler inlet and outlet enthalpy.

   At 4.5 MPa, 30 °C, h_1 = 129.75 kJ/kg

   At 4.5 MPa, 480 °C, h_2 = 3393.35 kJ/kg

   Heat absorption in the boiler:

   $$Q = \dot{m}(h_2 - h_1) = 20 \times 10^3 \times (3{,}393.35 - 129.75) = 6.5272 \times 10^7 \text{ kJ/h}$$

   Coal demand:

   $$\dot{m} = \frac{Q/0.8}{30{,}000} = \frac{65{,}272 \times 10^7/0.8}{30{,}000} = 2.7 \times 10^3 \text{ kg/h} = 2.7 \text{ t/h}$$

3. First check the steam table or diagram to determine the enthalpy and entropy.

   At 3 MPa, 350 °C, s_1 = 6.7443 kJ/(kg · K), h_1 = 3115.2 kJ/kg

   And $s_2 = s_1$ = 6.7443 kJ/(kg · K), s' = 0.5209 kJ/(kg · K), s'' = 8.3305 kJ/(kg · K)

   $$h' = 151.5 \text{ kJ/kg}, \quad h'' = 2{,}567.1 \text{ kJ/kg}$$

   $$x_2 = \frac{s_2 - s'}{s'' - s'} = \frac{6.7443 - 0.5209}{8.3305 - 0.5209} = 0.7969$$

   $$h_2 = x_2 h'' + (1 - x_2)h' = 0.7969 \times 2{,}567.1 + (1 - 0.7969) \times 151.5$$
   $$= 2{,}076.4 \text{ kJ/kg}$$

   $$N = \dot{m}|w_s| = \dot{m}(h_1 - h_2) = 3.4 \times (3{,}115.2 - 2{,}076.4) = 3{,}534.92 \text{ kW}$$

4. At 4 MPa, 450 °C, s_1 = 6.9379 kJ/(kg · K), h_1 = 3,330.7 kJ/kg

   According to p_2 = 0.005 MPa,

   $$s' = 0.4762 \text{ kJ/}(kg \cdot K), \quad s'' = 8.3952 \text{ kJ/}(kg \cdot K)$$

   $$h'_2 = 137.7 \text{ kJ/kg}, \qquad h''_2 = 2{,}561.2 \text{ kJ/kg}$$

   Determine the exit point 2 of the isentropic process, and the state parameters are given as follows, and $s_1 = s_2$:

   $$x_{2s} = \frac{s_2 - s'}{s'' - s'} = \frac{6.9379 - 0.4762}{8.3952 - 0.4762} = 0.816$$

   $$h_{2s} = x_{2s} h''_2 + (1 - x_{2s})h' = 0.816 \times 2{,}561.2 + (1 - 0.816) \times 137.77$$
   $$= 2{,}115.2 \text{ kJ/kg}$$

   And the actual exit point is 2 points when x = 0.9,

   $$h_2 = x_2 h''_2 + (1 - x_2)h' = 0.9 \times 2{,}561.2 + (1 - 0.9) \times 137.77 = 2{,}318.9 \text{ kJ/kg}$$

Internal efficiency of steam turbine:

$$\eta = \frac{h_1 - h_2}{h_1 - h_{2s}} = \frac{3{,}330.7 - 2{,}318.9}{3{,}330.7 - 2{,}115.2} = 83.2\%$$

5. Absolute pressure of exhausted steam: $p = p_b - p_4 = 0.1{-}0.094 = 0.006$ MPa, According to $p = 0.006$ MPa,

$$v' = 0.0010064 \, \text{m}^3/\text{kg}, \quad v'' = 23.742 \, \text{m}^3/\text{kg}$$

$$h' = 151.1 \, \text{kJ}/\text{kg}, \quad h'' = 2{,}567.1 \, \text{kJ}/\text{kg}$$

When $x = 0.9$:

$$v = xv'' + (1-x)v' = 21.362 \, \text{m}^3/\text{kg}$$

$$h = xh'' + (1-x)h' = 2{,}325.11 \, \text{kJ}/\text{kg}$$

So the size is reduced:

$$\frac{v}{v'} = \frac{21.362}{0.0010064} = 21{,}226$$

One kilogram of exhausted steam condenses into water to release heat: $q = h - h' = 2{,}174$ kJ.

6. According to the analysis, the process in the container is isovolumic: At 0.1 MPa, 20 °C, $v_1 = 1.0017 \times 10^{-3}$ m^3 / kg, $h_1 = 84$ kJ/kg According to $t_2 = 40$ °C, $v_2 = v_1 = 1.0017 \times 10^{-3}$ m^3/kg

$$p_2 = 14 \, \text{MPa}, \quad h_2 = 179.8 \, \text{kJ}/\text{kg}$$

$$Q = m(u_2 - u_1)$$

$$= m \times [(h_2 - p_2 v_2) - (h_1 - p_1 v_1)]$$

$$= m \times [(h_2 - h_1) - v(p_2 - p_1)]$$

$$= 20 \times \left[(179.8 - 84) - 1.0017 \times 10^{-3}(14 - 0.1) \times 10^3\right]$$

$$= 1{,}638 \, \text{kJ}$$

In addition, the heat added is calculated according to the fixed specific heat capacity:

$$Q = mc\Delta t = 20 \times 4.18 \times (40 - 20) = 1{,}672 \, \text{kJ}$$

Chapter 7
Steam and gas power cycles

7.1 Review: power cycle

Steam power cycle: This is the working cycle of steam engine and steam turbine, and working medium is steam, which is treated as the actual gas.

Gas power cycle: This is the working cycle of internal combustion engine and gas turbine, and the working medium is gas, which can be treated as an ideal gas.

The purpose of research on heat engine cycle:
(1) Analysis of the heat energy utilization economy (such as thermal efficiency)
(2) Exploration of factors causing thermal efficiency
(3) Ways to improve thermal efficiency

Research method of heat engine cycle:
(1) Establish a simplified thermodynamic model of the actual cycle
(2) Use simple and typical reversible processes and cycles to approximate the actual complex irreversible processes and cycles
(3) Determine the basic law through thermodynamic analysis

7.2 Steam power plant cycles

7.2.1 Rankine cycle

Steam power plant is a complete set of thermal equipment, including boiler, steam turbine, condenser, and water pump, which can be used in thermal power plants. The Rankine cycle is the simplest and most basic ideal steam power cycle obtained by simplification of the actual steam power cycle. It is the foundation of the study for other complex steam power cycles, as shown in Fig. 7.1.

Steam power cycle is the working cycle of steam power plant with water vapor as working medium, as shown in Fig. 7.2. The working cycle process can be simplified to as follows:

1–2: The superheated steam expands in the steam turbine to do work.

2–3: The low-pressure steam (exhaust gas) after work is condensed in the condenser to release heat.

3–4: The condensed water is compressed by the water pump and sent to the boiler.

4–1: The boiler supplies the energy required to vaporize the water and form high-temperature and high-pressure state.

https://doi.org/10.1515/9783111329703-007

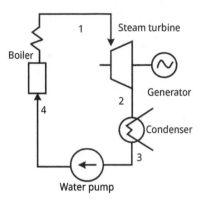

Fig. 7.1: Rankine cycle.

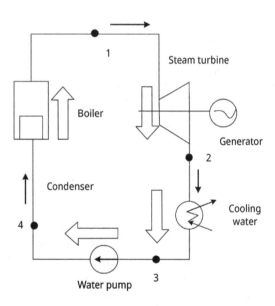

Fig. 7.2: Simplified ideal steam power cycle.

A simplified ideal steam power cycle consists of four idealized reversible processes, as shown in Fig. 7.3 and Fig. 7.4:

3–4: Reversible adiabatic compression process of water in feed pump

4–1: Reversible constant pressure heating process of water and steam in boiler

1–2: Reversible adiabatic expansion of water vapor in steam turbine

2–3: Reversible constant pressure heat release process of exhaust steam in condenser

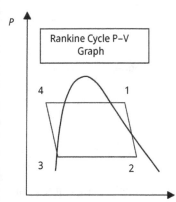

Fig. 7.3: *p–v* diagram of Rankine cycle.

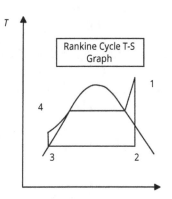

Fig. 7.4: *T–s* diagram of Rankine cycle.

7.2.2 Net work and thermal efficiency of Rankine cycle

Turbine work:

$$w_{s,1-2} = h_1 - h_2 \tag{7.1}$$

Heat release in condenser at constant pressure:

$$q_2 = h_2 - h_3 \tag{7.2}$$

Adiabatic compression power consumption of water pump:

$$w_{s,3-4} = h_4 - h_3 \tag{7.3}$$

Heat absorption in boiler at constant pressure:

$$q_1 = h_1 - h_4 \tag{7.4}$$

Net work per kg of steam:

$$w_{net} = (h_1 - h_2) - (h_4 - h_3) \tag{7.5}$$

The constant pressure heat absorption per kg of steam in the boiler:

$$q_1 = h_1 - h_4 \tag{7.6}$$

The thermal efficiency of the Rankine cycle is defined as

$$\eta_t = \frac{w_{net}}{q_1} = \frac{w_{s,1-2} - w_{s,3-4}}{q_1} = \frac{(h_1 - h_2) - (h_4 - h_3)}{h_1 - h_4} \tag{7.7}$$

Due to the small compressibility of the water, the work consumed by the water pump is very small compared with the work made by the turbine, which can be ignored:

$$h_4 - h_3 \approx 0 \tag{7.8}$$

$$\eta_t = \frac{h_1 - h_2}{h_1 - h_4} = \frac{h_1 - h_2}{h_1 - h_3} \tag{7.9}$$

Concept of steam rate: The steam quantity of kg consumed by the steam power plant for each 1 kW·h power output is in units of kg/(kW·h), which reflects the plant economy and equipment size in engineering:

$$d = \frac{3,600}{w_{net}} \tag{7.10}$$

Example: The parameters of the new steam in a Rankine cycle are $p_1 = 5$ MPa, $t_1 = 550$ °C, and the pressure of the exhaust gas $p_2 = 5$ kPa. Ignore the pump work, and try to calculate the network of the cycle, heat addition, thermal efficiency, and the dryness of the exhaust gas x.

First, the cycle is represented on the T–S diagram, as shown in Fig. 7.5.

Point 1: $p_1 = 5$ MPa, $t_1 = 550$ °C. Table lookup: $h_1 = 3548.0$ kJ/kg, $s_1 = 7.1187$ kJ/(kg·K)

Point 2: $s_2 = s_1 = 7.1187$ kJ/(kg·K). According to s_2 and p_2, x can be calculated.

For the saturated steam and water corresponding to "2" point, from $p_2 = 5$ kPa, look up the table and get s_2' and s_2'', h_2' and h_2''.

Point 3(4): According to $p_3 = p_2 = 5$ kPa, table lookup: h_3, heat absorption: $q_1 = h_1-h_3$, and network: $w_{net} = h_1-h_2$:

$$x = \frac{s_2 - s_2'}{s_2'' - s_2'} \tag{7.11}$$

$$h_2 = xh_2'' + (1-x)h_2' \tag{7.12}$$

$$\eta = \frac{w_{net}}{q_1} \tag{7.13}$$

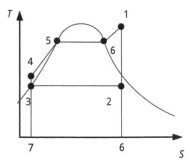

Fig. 7.5: T–s diagram.

7.2.3 Effects of steam parameters on Rankine cycle thermal efficiency

The formula for thermal efficiency is as follows:

$$\eta_t \approx \frac{h_1 - h_2}{h_1 - h_4} \approx \frac{h_1 - h_2}{h_1 - h_3} \tag{7.14}$$

There are also some concepts that need to be explained:
(1) Temperature t_1 of new steam (initial temperature)
(2) Pressure of new steam p_1 (initial pressure)
(3) Pressure of exhaust steam p_2 (final pressure)

The Rankine cycle can be converted into a Carnot cycle with the same entropy, heat absorption (emission), and thermal efficiency, as shown in Fig. 7.6.

Average temperature of endothermic:

$$\bar{T}_1 = \frac{q_1}{s_a - s_b} \tag{7.15}$$

Average temperature of exothermic:

$$\bar{T}_2 = T_2 \tag{7.16}$$

$$\eta_t = \frac{h_1 - h_2}{h_1 - h_3} = 1 - \frac{T_2}{\bar{T}_1} \tag{7.17}$$

The methods to improve the thermal efficiency of the cycle are:
Increasing the average temperature of heat absorption (T_1) and decreasing the average temperature of heat release (T_2).

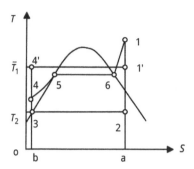

Fig. 7.6: *T–s* diagram.

7.2.3.1 The effect of initial steam pressure

When t_1 and p_2 remain unchanged and p_1 increased, as shown in Fig. 7.7(a):

The advantage is that the average temperature of endothermic $\overline{T_1}$ is increased and the thermal efficiency η_t is increased. At the same time, due to the reduction of the specific volume v_2' of the exhausted steam, the size of the turbine outlet is reduced, and the economy is better.

But with the increase of the initial pressure p_1, the dryness of the exhausted steam x_2 decreases, which means that the water content in the exhausted steam increases. When it exceeds a certain limit, it will impact and erode the last stages of the turbine blades, affecting its service life. At the same time, the friction loss inside the steam turbine is increased, and the working effect is reduced. Therefore, in engineering, the initial temperature is usually increased while the initial pressure is increased to ensure that the dryness of the spent steam is not less than 0.85–0.88.

7.2.3.2 The effect of steam initial temperature

When p_1 and p_2 remain unchanged, t_1 increased, as shown in Fig. 7.7(b).

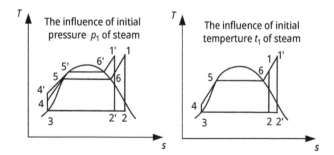

Fig. 7.7: Effect of initial steam pressure.

The advantage is that the average temperature of endothermic $\overline{T_1}$ is increased, the thermal efficiency η_t is increased, and x is also increased. And it is also beneficial to the safe work of the steam turbine.

However, the disadvantage is that due to the increase of the initial temperature of the steam, the heat resistance and strength of the metal are required to be high, so the initial temperature is generally set at about 550 °C. And because the specific volume v_2 of the exhausted steam at the steam turbine outlet becomes larger, the outlet size of the large turbine also becomes larger.

7.2.3.3 The effect of exhaust steam pressure

When p_1 and t_1 remain unchanged, p_2 increased, as shown in Fig. 7.8.

Because the initial steam parameters are kept unchanged, the exhausted steam pressure p_2 is reduced. The corresponding saturation temperature $\overline{T_2}$ (exothermic temperature) decreases, while the average endothermic temperature changes little, so the thermal cycle efficiency η_t will be improved.

However, the reduction of the final pressure p_2 is limited by the temperature of cooling medium (environment temperature) and cannot be reduced arbitrarily. p_2 can only be reduced to a minimum of 0.0035–0.005 MPa, and the corresponding saturation temperature is about 27–33 °C, which is close to the minimum possible.

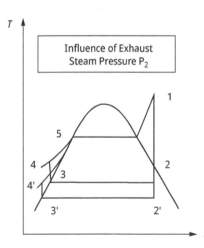

Fig. 7.8: Effect of exhaust steam pressure in T–s diagram.

In summary, in order to improve the thermal efficiency of steam power cycle, the initial pressure and temperature of steam should be elevated as much as possible, and the pressure of exhaust steam should be reduced.

7.2.4 Other ways to improve the thermal efficiency of steam power cycle

Change cycle parameters: Increase initial temperature, increase initial pressure, and reduce exhaust steam pressure
Change the cycle form: Reheat cycle and regenerative cycle
Types of combined cycle: Cogeneration cycle, gas steam combined cycle, and new power cycle

7.2.4.1 Reheat cycle
The new steam expands and works to a certain pressure in the high-pressure steam turbine, all of which are drawn out and led into the boiler for reheating, and then led to the low-pressure steam turbine to expand and work to p_2. This is shown in Fig. 7.9.

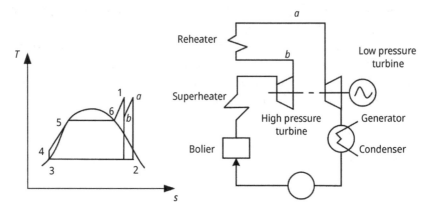

Fig. 7.9: Ideal reheat cycle.

However, the reheat cycle itself does not necessarily improve the thermal efficiency of the cycle. The increase in thermal efficiency is related to the reheat pressure.

After the steam is reheated in the middle, the dryness x_2 of the spent steam is obviously improved. The increase of x_2 creates conditions for the increase of initial pressure, and the selection of reheat pressure is appropriate. Generally, one reheat can increase the thermal efficiency by 2–3.5%.

Heat absorption:

$$q_1 = (h_1 - h_4) + (h_a - h_b) \tag{7.18}$$

Heat release:

$$q_2 = h_2 - h_3 \tag{7.19}$$

Network (ignoring pump power):

$$w_{net} = (h_1 - h_b) - (h_a - h_3) \tag{7.20}$$

Thermal efficiency:

$$\eta_{t,RH} = \frac{w_{net}}{q_1} = \frac{(h_1 - h_b) + (h_a - h_2)}{(h_1 - h_4) + (h_a - h_b)} \tag{7.21}$$

7.2.4.2 Regenerative vapor power cycle

A small amount of steam from the middle part of the turbine that has done work but the pressure is not too low is used to heat the low-temperature feed water before entering the boiler. This is shown in Fig. 7.10. The left side of Fig. 7.11 is a schematic diagram of a steam regeneration cycle system for primary steam extraction, and the right side of Fig. 7.11 is a T–s diagram of the regeneration cycle. One kilogram of fresh steam with pressure p_1 enters the steam turbine to expand and do work, and the state changes from 1 to 7. At this time, c kg ($c < 1$) steam is extracted and introduced into the regenerative heater, where it condenses and releases heat along the process line 7–8–9, and the remaining $(1-c)$ kg steam continues to expand in the steam turbine to do work until the exhausted steam pressure is p_2. Then it enters the condenser and is condensed into water, which is boosted by the condensate pump and enters the regenerative heater, receives the latent heat released during the condensation of c kg steam extraction and mixes with it to become 1 kg saturated water under the steam extraction pressure, and finally passes through the water pump. This is pressurized into the boiler to absorb heat, vaporize, and superheat to become new steam to complete a cycle.

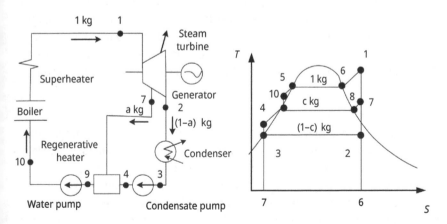

Fig. 7.10: Regenerative vapor power cycle.

The average temperature of heat absorption is increased, and the thermal efficiency of circulation is improved (the temperature of boiler feed water is increased from T_4 to T_{10}).

According to the first law of thermodynamics, the heat released by a kg extraction should be equal to the saturated temperature of $(1 - a)$ kg condensate heated to extraction pressure. This is shown in Fig. 7.11:

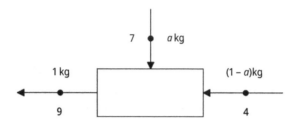

Fig. 7.11: **Example of** regenerative vapor.

According to the first law of thermodynamics, the energy balance in the regenerative heater is as follows:

$$a(h_7 - h_9) = (1 - a)(h_9 - h_4) \tag{7.22}$$

$$a = \frac{h_9 - h_4}{h_7 - h_4} \approx \frac{h_9 - h_3}{h_7 - h_3} \tag{7.23}$$

$$\eta_t = 1 - \frac{q_2}{q_1} = 1 - \frac{(1 - a)(h_2 - h_3)}{h_1 - h_9} \tag{7.24}$$

The selection of extraction pressure is a problem that must be considered. It depends on the temperature of the feed water before the boiler ($t_{10} \approx t_9$). It will not achieve the purpose of improving the thermal efficiency of the cycle when temperature is too high or too low.

Therefore, the heat recovery stage numbers of small thermal power plants are generally 1–3, and that of medium and large thermal power plants are generally 4–8.

7.2.4.3 Cogeneration cycle

In order to make full use of thermal energy, the waste heat of exhaust steam in power plant is used to meet the needs of heat users. A power plant that provides both electricity and heat is customarily called a thermal power plant, and is shown in Fig. 7.12.

The working fluid absorbs heat in the working part and the heating part.

In the cogeneration cycle, the heat absorbed by the working medium is not only used for power generation but also provides heat to users. Therefore, in addition to the thermal efficiency η_t, the thermal economic index for evaluating the cogeneration cycle also has the energy utilization coefficient:

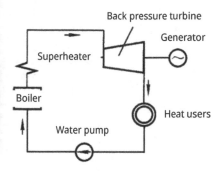 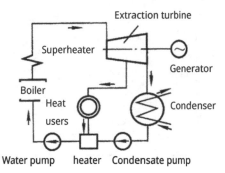

Fig. 7.12: Cogeneration cycle.

$$K = \frac{q_{supply} - W_{net}}{q_1} \tag{7.25}$$

For back pressure cogeneration cycle, theoretically K can reach 1, but due to various losses and leaks, K value is only about 0.85.

7.3 Piston internal combustion engine cycle

Taking a four-stroke diesel engine as an example to analyze its actual working cycle, the internal combustion engine has four strokes in each working cycle; for example, the piston of each cycle goes back and forth twice in the cylinder. The four strokes are intake stroke, compression stroke, power stroke, and exhaust stroke [13].

7.3.1 Piston engine actual cycle

The four strokes of the working cycle of the diesel engine are as follows. The diagram of piston engine's actual cycle is shown in Fig. 7.13.

Intake stroke 0–1: The piston moves down from the top dead center of the cylinder, the intake valve opens, and the air is drawn in. Due to the throttling effect of the intake valve, the pressure of the gas in the cylinder is slightly lower than the atmospheric pressure.

Compression stroke 1–2: When the piston reaches the lower dead center point 1, the inlet valve closes, the piston goes up, and the air is compressed.

1–2′ is a polytropic compression process, and the specific parameters at 2′ are: $p_2' = 3$–5 MPa, $t_2' = 600$–800 °C. Diesel fuel injection is at time 2′, and starts burning at time 2. Diesel spontaneous combustion $t = 205$ °C.

Power stroke 2–3–4–5: 2–3 Rapid combustion with approximately constant volume V and p increased to 5–9 MPa.

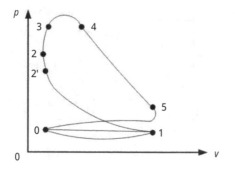

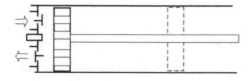

Fig. 7.13: Piston engine's actual cycle.

The 3–4 process is fuel injection and expansion, with approximate isobaric expansion (constant p), where t_4 can reach 1,700–1,800 °C. Time 4 is to stop diesel injection, and the process of 4–5 is polytropic expansion, where $p_5 = 0.3$–0.5MPa, $t_5 \approx 500$ °C.

Exhaust stroke 5–0: Open the valve to exhaust and reduce the pressure to slightly higher than the atmospheric pressure at constant volume V. Then the piston goes up to exhaust and completes the cycle.

7.3.2 Ideal cycle of piston engine

Abstract, summarize, and simplify the actual cycle reasonably [14]:
1) Ignoring the throttling loss of intake and exhaust valves in the actual process, the intake process and exhaust process cancel each other, and open to closed cycle.
2) Assume that the working medium is an ideal gas with constant chemical composition and constant specific heat capacity – air.
3) It is considered that the expansion and compression process of the working medium is reversible and adiabatic.
4) The combustion process of fuel is regarded as the reversible heat absorption process of working medium from high-temperature heat source, and the exhaust process is regarded as the reversible heat release process of working medium to low-temperature heat source.
5) Ignore the dynamic and potential energy changes of working medium.

After the above simplification, abstraction, and generalization, the actual diesel engine cycle can be idealized as an ideal reversible cycle as shown in Fig. 7.14.

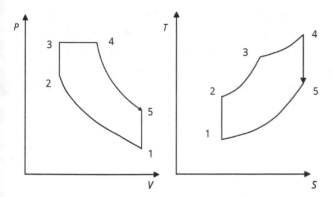

Fig. 7.14: Ideal cycle of piston engine.

The specific process of the ideal cycle of piston engine is as follows:

1–2: Adiabatic compression

2–3: Constant volume heating

3–4: Constant pressure heating

4–5: Adiabatic expansion

5–1: Constant volume heat release

This cycle is called mixed heating cycle, also known as the mixed cycle (Sabathe cycle).

Self-study work heat calculation:

$$\eta_{t,m} = 1 - \frac{\lambda \rho^k - 1}{\varepsilon^{k-1}[\lambda - 1 + k\lambda(\rho - 1)]} \tag{7.26}$$

7.4 Ideal cycle of gas turbine plant

The compression, combustion, and expansion of piston internal combustion engine are carried out in the same cylinder in sequence and repeatedly. The power of the engine is limited by the discontinuity of air flow and the influence of inertia force on rotating speed when piston reciprocates.

If the compressor, combustion, and expansion are, respectively, carried out in the compressor, combustion chamber, and gas turbine, a new type of internal combustion power plant will be formed – gas turbine plant.

7.4.1 Open-type gas turbine (actual situation)

An open-type gas turbine consists of three components: the compressor, the combustor, and the gas turbine. Open gas turbines work by taking in fresh atmospheric air and compressing it using compressor. Compressed air is injected into the combustion chamber, where it mixes with the fuel, producing hot gases at high pressure. The high-pressure hot gas is used to drive the turbine, and as the gas passes through the turbine blades, it turns a shaft connected to the rotor portion of the generator. The power generated by turning the turbine shaft can power a variety of industrial equipment and be used to generate electricity. This is shown in Fig. 7.15.

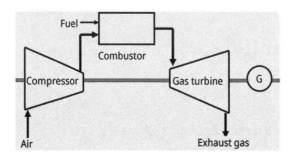

Fig. 7.15: Open-type gas turbine.

7.4.2 Ideal closed gas turbine

If a cooler which is using to cool water is added to the above open gas turbine system, the whole system will become a closed gas turbine. This is shown in Fig. 7.16.

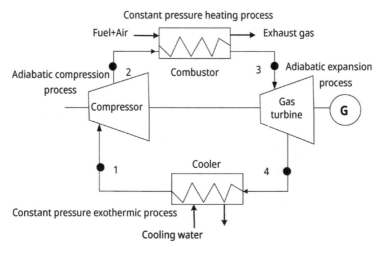

Fig. 7.16: Closed gas turbine.

In order to perform a thermodynamic analysis of the cycle, the closed gas turbine is idealized:

(1) Assume that the working fluid is an ideal gas with a constant specific heat capac-
     ity – air, ignoring the mass of the injected fuel;
(2) All processes experienced by the working fluid are reversible processes;
(3) In the compressor and the gas turbine, the processes experienced by the working
     fluid are all adiabatic processes;
(4) The working fluid in the combustion chamber experiences a constant pressure
     heating process;
(5) The exothermic process of the working medium to the atmosphere is a constant
     pressure exothermic process.

Figure 7.17 shows the p–v and T–s diagrams of the ideal cycle described above. In this
figure, 1–2 is the reversible adiabatic compression process of air in the compressor;
2–3 is the reversible constant pressure heating process of air in the combustion chamber; 3–4 is the reversible adiabatic expansion process of air in the gas turbine; 4–1 is
the reversible constant pressure exothermic process of air in the atmosphere.

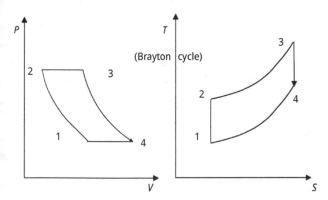

Fig. 7.17: Constant pressure heating gas turbine unit cycle.

Heat absorption:

$$q_1 = c_p(T_3 - T_2) \tag{7.27}$$

Heat release:

$$q_2 = c_p(T_4 - T_1) \tag{7.28}$$

Thermal efficiency:

$$\eta_t = \frac{w}{q_1} = \frac{q_1 - q_2}{q_1} = 1 - \frac{q_2}{q_1} = 1 - \frac{T_4 - T_1}{T_3 - T_2} \tag{7.29}$$

$$\frac{T_3}{T_4} = \left(\frac{p_3}{p_4}\right)^{\frac{k-1}{k}} = \left(\frac{p_2}{p_1}\right)^{\frac{k-1}{k}} = \frac{T_2}{T_1} \tag{7.30}$$

$$\eta_t = 1 - \frac{T_1}{T_2} = 1 - \frac{1}{\frac{T_2}{T_1}} = 1 - \frac{1}{\left(\frac{p_2}{p_1}\right)^{\frac{k-1}{k}}} \tag{7.31}$$

Pressure ratio:

$$\pi = \frac{p_2}{p_1} \tag{7.32}$$

$$\eta_t = 1 - \frac{1}{\pi^{\frac{k-1}{k}}} \tag{7.33}$$

7.5 Summary

(1) Steam power cycle: master the definition and working process of Rankine cycle; understand the influencing factors of Rankine cycle efficiency and the ways to improve the cycle efficiency.

(2) Piston internal combustion engine cycle: understand the working principle and the composition of the cycle.

(3) Gas power plant cycle: understand the working principle, cycle composition, and calculation of cycle efficiency.

Exercises

1. In a Rankine cycle, the parameters of new steam are p_1, t_1, and the pressure of exhaust steam p_2. Ignore the pump work, and try adding the appropriate parameters to calculate the net circulating work of this cycle, heating capacity, thermal efficiency, and the dryness x of exhaust steam.

2. The inlet gas state of a gas turbine is $p_1 = 0.1$ MPa, $t_1 = 22$ °C, the cycle pressure ratio $\pi = 7$, the working medium is air, the specific heat capacity is constant, and the temperature after heat absorption at constant pressure is 600 °C. Try to calculate the shaft work consumed by the compressor, the shaft work done by the gas turbine, the net work output by the gas turbine, and the circulating thermal efficiency.

3. The initial temperature of the steam Rankine cycle is $t_1 = 500$ °C, the back pressure (depleted steam pressure) $p_2 = 0.004$ MPa; ignoring the pump power, try to find the circulating network, heating capacity, thermal efficiency, steam consumption rate, and steam turbine outlet dryness x_2 when the initial pressure $p_1 = 4$ MPa.

4. The initial temperature of a steam power cycle is $t_1 = 380$ °C, the initial pressure $p_1 = 2.6$ MPa, and the back pressure $p_2 = 0.07$ MPa. If the relative internal efficiency of the steam turbine is $\eta_{oi} = 0.8$, the pump work is ignored. Find the cycle's specific work, thermal efficiency, and steam consumption rate.

Answers

1.
$$P_1, t_1 \rightarrow h_1, s_1 s_2 = s_1$$
$$P_2 \rightarrow s_2', s_2'', h_2', h_2''$$
$$X = \frac{s_2 - s_2'}{s_2'' - s_2'}$$
$$h_2 = xh_2'' + (1-x)h_2'$$
$$h_3 = h_2'$$
$$q_1 = h_1 - h_3$$
$$w_{net} = h_1 - h_2$$
$$\eta = \frac{w_{net}}{q_1}$$

2.
$$T_2 = T_1 \pi^{\frac{k-1}{k}} = 295 \times 7^{\frac{1.4-1}{1.4}} = 514.37 \text{ K}$$

$$T_4 = T_3/\pi^{\frac{k-1}{k}} = 873/7^{\frac{1.4-1}{1.4}} = 500.68 \text{ K}$$

$$w_1 = h_2 - h_1 = c_p(T_2 - T_1)$$
$$= 1{,}004.5 \times (514.37 - 295) = 220.36 \text{ kJ/kg}$$

$$w_2 = h_3 - h_4 = c_p(T_3 - T_4)$$
$$= 1{,}004.5 \times (873 - 500.68) = 374.00 \text{ kJ/kg}$$

$$w_{net} = w_2 - w_1 = 374.00 - 220.36 = 153.64 \text{ kJ/kg}$$

$$q_1 = h_3 - h_2 = c_p(T_3 - T_2)$$
$$= 1{,}004.5 \times (873 - 514.37) = 360.24 \text{ kJ/kg}$$

$$\eta = \frac{w_{net}}{q_1} = \frac{153.64}{360.24} = 42.65\%$$

$$\eta = 1 - \frac{1}{\pi^{\frac{k-1}{k}}} = 1 - \frac{1}{\pi^{\frac{1.4-1}{1.4}}} = 42.65\%$$

3. Ignoring pump work, the T–s diagram of the Rankine cycle is shown in Fig. 7.18.

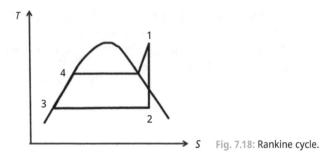

Fig. 7.18: Rankine cycle.

From $p_1 = 4$ MPa, $t_1 = 500$ °C, we know:
$h_1 = 3,445.2$ kJ/kg, $s_1 = 7.0909$ kJ/(kg · K)
From $p_2 = 0.004$ MPa, we find:
$h_2' = 121.41$ kJ/kg, $h_2'' = 2,554.1$ kJ/kg
$s_2' = 0.4224$ kJ/(kg · K), $s_2'' = 8.4747$ kJ/(kg · K)
Because of equal dilatation, $s_2 = s_1$
Therefore,

$$x_2 = \frac{s_1 - s_2'}{s_2'' - s_2'} = 0.828$$

$$h_2 = x_2 h_2'' + h_2'(1 - x_2) = 2,135.68 \text{ kJ/(kg · K)}$$

$$w = h_1 - h_2 = 1,309.52 \text{ kJ/kg}$$

$$d = \frac{3,600}{w} = 2.75 \text{ kg/(kW · K)}$$

$$q_1 = h_1 - h_2' = 3,323.78 \text{ kJ/kg}$$

$$\eta_t = \frac{w}{q_1} = 39.4\%$$

4. From $p_1 = 2.6$ MPa, $t_1 = 380$ °C, we know:
    $h_1 = 3,193.08$ kJ/kg, $s_1 = 6.9373$ kJ/(kg · K)
    From $p_2 = 0.007$ MPa, we find:
    $h_2' = 163.38$ kJ/kg, $h_2'' = 2,572.2$ kJ/kg
    $s_2' = 0.5591$ kJ/(kg · K), $s_2'' = 8.276$ kJ/(kg · K), $s_2 = s_1 = 6.9373$ kJ/(kg · K)

$$x_2 = \frac{s_2 - s_2'}{s_2'' - s_2'} = \frac{6.9373 - 0.5591}{8.276 - 0.5591} = 0.8265$$

$$h_2 = h_2' + x_2\left(h_2'' - h_2'\right)$$

$$= 163.38 + 0.8265 \times (2,572.2 - 163.38) = 2,154.27 \text{ kJ/kg}$$

$$h_2' = h_1 - \eta_{oi}(h_1 - h_2)$$
$$= 3{,}193.08 - 0.8 \times (3{,}193.08 - 2{,}154.27) = 2{,}362 \ \text{kJ/kg}$$

$$x_{2'} = \frac{h_{2'} - h_2'}{h_2'' - h_2'} = \frac{2{,}362 - 163.38}{2{,}572.2 - 163.38} = 0.9127$$

$$w'_{net} = h_1 - h_{2'} = 831.08 \ \text{kJ/kg}$$

$$\eta_t' = \frac{w'_{net}}{q_1} = \frac{831.08}{3{,}193.08 - 163.38} = 27.4\%$$

$$d = \frac{3{,}600}{w'_{net}} = 4.33 \ \text{kg/(kW} \cdot \text{K)}$$

Chapter 8
Refrigeration equipment and cycle

8.1 Refrigeration

Refrigeration is using external work to transfer heat continuously from low-temperature object to high-temperature environment medium. Equipment that can maintain a cold region at low temperature is called a refrigeration device.

The refrigeration plant cycle is a reverse cycle according to the second law of thermodynamics. In order to obtain and maintain the low temperature, the equipment achieves the goal of "taking heat away from the cold object or cooling space" at the cost of energy consumption (work or heat).

Compression refrigeration devices are currently widely used refrigeration devices, and most household refrigerators are air conditioners and freezers. If the refrigerant is always in a gaseous state during the cycle, the refrigeration cycle is called a gas compression refrigeration cycle. If the state of the refrigerant changes across the liquid and gas states, the refrigeration cycle is called a vapor compression refrigeration cycle. The refrigeration method can be classified as gas compression refrigeration cycle, vapor compression refrigeration cycle, absorption refrigeration cycle, adsorption refrigeration cycle, steam jet refrigeration cycle, semiconductor, and so on.

This chapter mainly introduces the working principles of two compression refrigeration cycles, absorption refrigeration cycles, and heat pumps [15].

8.2 Air compression refrigeration cycle

8.2.1 The principle of air compression refrigeration cycle

The principle of air compression refrigeration cycle is that it uses the cooling of the air as it expands adiabatically to obtain low temperatures.

The schematic diagram of the air compression refrigeration cycle device is shown in Fig. 8.1. The air coming out of the heat exchanger in the cold chamber is sucked in and compressed by the compressor. After increasing the pressure and temperature, the air enters the cooler. After being cooled, it enters the expander to expand and do work, and the pressure and temperature drop significantly. The low-temperature and low-pressure air enters the refrigerating room to absorb heat, thereby achieving the purpose of maintaining the low temperature of the cold chamber (i.e., refrigeration). The air after endothermic heating is sucked into the compressor again for the next cycle [16].

https://doi.org/10.1515/9783111329703-008

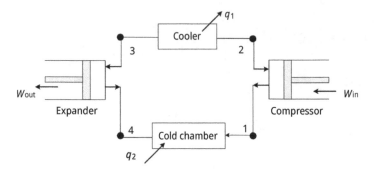

Fig. 8.1: Air compression refrigeration cycle.

8.2.2 *p–v* diagram and *T–s* diagram of air compression refrigeration cycle

If the actual work of an air compression refrigeration device is idealized, the gas should be regarded as an ideal gas, the specific heat of the process should be constant, and the process should be reversible. Then the above process is: 3–4 adiabatic expansion, 4–1 constant pressure heating, 1–2 adiabatic compression, and 2–3 constant pressure cooling. The *p–v* diagram and *T–s* diagram of the air compression refrigeration cycle are shown in Fig. 8.2.

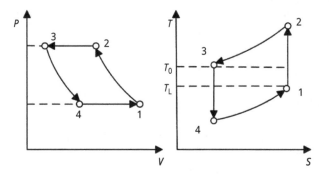

Fig. 8.2: *p–v* diagram and *T–s* diagram of air compression refrigeration cycle.

If the specific heat capacity of air is constant, the heat released per unit mass air in the cooler is

$$q_1 = h_2 - h_3 = c_p(T_2 - T_3) \tag{8.1}$$

In the cold chamber, the heat absorbed by the air from low-temperature objects is

$$q_2 = h_1 - h_4 = c_p(T_1 - T_4) \tag{8.2}$$

$$\Delta h = \int_1^2 c_p dT \tag{8.3}$$

For reversible adiabatic processes 1–2 and 3–4:

$$\frac{T_2}{T_1} = \left(\frac{p_2}{p_1}\right)^{\frac{\kappa-1}{\kappa}} \tag{8.4}$$

$$\frac{T_3}{T_4} = \left(\frac{p_3}{p_4}\right)^{\frac{\kappa-1}{\kappa}} \tag{8.5}$$

Thus, the refrigeration coefficient of the air-compressed refrigeration cycle is

$$\varepsilon = \frac{q_2}{q_1 - q_2} \tag{8.6}$$

Due to $p_2 = p_3$, $p_1 = p_4$:

$$\frac{T_2}{T_1} = \frac{T_3}{T_4} = \frac{T_2 - T_3}{T_1 - T_4} = \left(\frac{p_2}{p_1}\right)^{\frac{\kappa-1}{\kappa}} = \pi^{\frac{\kappa-1}{\kappa}} \tag{8.7}$$

8.2.3 Regenerative air compression refrigeration unit

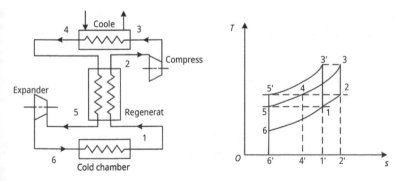

Fig. 8.3: Regenerative air compression refrigeration unit.

The right side of Fig. 8.3 is a schematic diagram of a regenerative air compression refrigeration device, and the left side of it is the T–s diagram of its ideal cycle. In this figure: process 1–2 is the constant pressure preheating process of air in the regenerator;

process 2–3 is the adiabatic compression process of air in the compressor; process 3–4 is the constant pressure heat release process of the air in the cooler; process 4–5 is the constant pressure heat release process of the air in the regenerator; process 5–6 is the adiabatic expansion process of air in the expander; and process 6–1 is the constant pressure endothermic process of air in the refrigerating room. This constitutes an ideal regenerative air compression refrigeration cycle 1234561. The non-regenerative cycle of 13'5'61 is compared with the regenerative cycle 1234561. Because it is an ideal regenerative process, the heat released by the air in process 4–5 is exactly equal to the heat absorbed by the air in the process 1–2, that is, the area of 456'4'4 is equal to the area of 122'1'1.

The cyclic cooling capacity q_2 and heat release q_1 of cycle 1234561 and cycle 123'5'61 are the same, which means that the refrigeration coefficient is the same. But the pressure ratio is reduced, and the impeller compressor and expander with small pressure ratio and large flow can be used to increase the flow of working medium and the total amount of refrigeration.

8.2.4 Disadvantages of air compression refrigeration cycle

(1) Unable to achieve constant temperature heat absorption and release process
(2) The cycle deviates from the reverse Carnot cycle, reducing the economy
(3) The cooling capacity of unit mass working medium is small
(4) The c_p of air is small

Regeneration only improves the refrigeration cycle, but it cannot eliminate the fact that the refrigeration capacity is small. Therefore, some solutions are proposed below:

The low boiling point material is used as a refrigerant, and its characteristic of constant pressure (constant temperature) in the wet vapor region is utilized to realize the constant temperature vaporization and endothermic refrigeration at low temperature.

8.3 Vapor compression refrigeration cycle

The schematic diagram of the air compression refrigeration cycle and the vapor compression refrigeration cycle is shown in Fig. 8.4. It can be seen from the figure that the number of devices in the two cycles is the same, but the arrangement order of the devices is different.

The working cycle of the vapor compression refrigeration device is as follows: the dry saturated vapor in state a from the evaporator is sucked into the compressor for adiabatic compression process a–b, and the pressure is increased and heated to the superheated vapor state b. The superheated steam enters the condenser and undergoes a constant pressure and heat release process b–c. It is cooled from the superheated steam state b under constant pressure to dry saturated steam b', and then

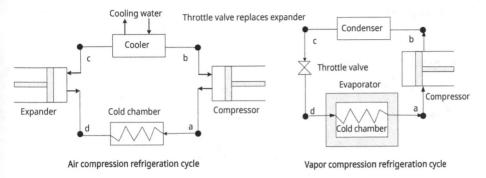

Fig. 8.4: Air compression refrigeration cycle and vapor compression refrigeration cycle.

continues to condense into saturated liquid c under constant pressure and temperature. The saturated liquid from the condenser passes through adiabatic throttling of
the expansion valve so that part of the liquid is evaporated, depressurized, and cooled
to the wet vapor at state d. The wet vapor with a relatively small dryness enters the
evaporator (cold chamber), and the constant pressure evaporation and heat absorption process are carried out at d–a. It leaves the evaporator as dry saturated vapor,
completing one cycle abcda. Figure 8.5 is the $T–s$ diagram of the ideal cycle of vapor
compression refrigeration, in which the adiabatic throttling process $c–d$ is an irreversible process, so it is represented by a dotted line.

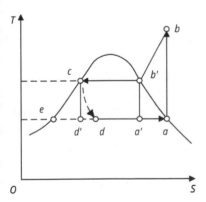

Fig. 8.5: $T–s$ graph of vapor compression refrigeration cycle.

a–b: Adiabatic compression process
b–c: Constant pressure condensation exothermic process
c–d: Adiabatic throttling process
d–a: Constant pressure evaporation endothermic process

Heat absorption:

$$q_2 = h_1 - h_4 = h_1 - h_3 \tag{8.8}$$

Heat release:

$$q_1 = h_2 - h_3 \tag{8.9}$$

Coefficient of refrigeration:

$$\varepsilon = \frac{q_2}{q_1 - q_2} = \frac{h_1 - h_3}{h_2 - h_1} \tag{8.10}$$

The pressure–enthalpy diagram of the compression refrigeration cycle is shown in Fig. 8.6.

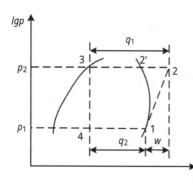

Fig. 8.6: Pressure–enthalpy diagram of compression refrigeration cycle.

The refrigerant characteristics are as follows:
(1) The saturation pressure corresponding to atmospheric temperature T_0 shall not be too high to reduce the equipment strength and sealing requirements.
(2) The saturation pressure corresponding to the cold storage temperature T_L should not be too low, and it is better not to operate under negative pressure.
(3) The latent heat of vaporization is large and the specific volume of vapor is small, so the refrigerating capacity per unit mass is large.
(4) The higher critical temperature makes most of the heat release in the two-phase region at a constant temperature.
(5) Low freezing point, low price, non-toxic, non-corrosive, non-explosive, stable in nature, non-polluting, and easily available.

8.4 Absorption refrigeration cycle

8.4.1 Characteristics of absorption refrigeration

Compression refrigeration is powered by work, while absorption refrigeration is powered by heat, and low-grade energy can be used.

The working pairs in absorption refrigeration cycle are binary solutions with big difference in boiling point, which are collectively called refrigerants as solutes and absorbers as solvents; different solution concentrations correspond to different saturation pressures and temperatures.

For example, in ammonia/hydraulic pair, ammonia is solute and water is solvent; in lithium bromide/hydraulic pair, water is solute and lithium bromide is solvent.

8.4.2 Absorption refrigeration equipment

In a vapor compression refrigeration device, the low-temperature and the low-pressure vapor coming out of the evaporator is compressed into a high-temperature and high-pressure superheated vapor by the compressor. Compression of the working fluid can also be achieved by other means. The compression method in the absorption refrigeration cycle is to use the characteristics of the solubility of the solute (refrigerant) in the solvent (absorbent) changing with temperature (the lower temperature, the higher solubility; the higher temperature, the lower solubility) characteristics so that the refrigeration working medium is at a lower temperature. It is absorbed by the absorbent to form a binary solution at a high temperature, and then escapes from the solution at a higher pressure after heating, thereby completing the compression process of the refrigerant. Figure 8.7 shows the ammonia–water absorption refrigeration device

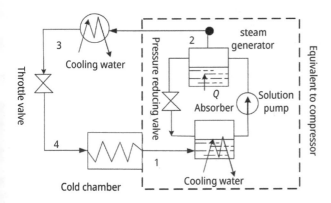

Fig. 8.7: Ammonia–water absorption refrigeration cycle.

Absorber: Dilute solution absorbs refrigerant with low pressure and temperature.

Steam generator: Concentrated solution releases refrigerant with high pressure and temperature.

Absorber, solution pump, steam generator, and pressure reducing valve are equivalent to the compressor of compression refrigeration, but the driving force is low-grade heat.

8.5 Heat pump

Heat pump device and refrigeration device work on the same principle, but the purpose is different: heat pump for heating and refrigeration device for refrigeration. Figure 8.8 shows the refrigeration cycle device in summer and the heat pump cycle device in winter [17].

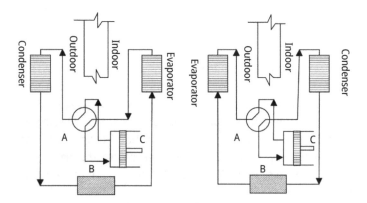

Fig. 8.8: Left: refrigeration cycle in summer. Right: heat pump cycle in winter.

In the figure, "A" represents the four-way reversing valve, "B" represents the capillary throttle device, and "C" represents the compressor.

Heating coefficient of a heat pump:

$$\varepsilon' = \frac{q_1}{w} \tag{8.11}$$

Refrigeration coefficient of refrigeration unit:

$$\varepsilon = \frac{q_2}{w} \tag{8.12}$$

The relationship between heating coefficient and refrigeration coefficient is

$$\varepsilon' = \frac{w + q_2}{w} = \varepsilon + 1 \tag{8.13}$$

A heat pump device can raise a large amount of low-grade (i.e., lower temperature) thermal energy to higher grade (i.e., higher temperature) thermal energy to meet the needs of production and life.

Exercises

1. In an inverse Carnot refrigeration cycle, the coefficient of refrigeration $\varepsilon = 5$, and what is the temperature ratio of the high-temperature heat source to the low-temperature heat source? If the input power is 6 kW, what is the cooling capacity? If it is used as a heat pump, what is the heating coefficient and the amount of heat provided?

2. A refrigerator works between 245 and 300 K, the heat absorption is 9 kW, and the refrigeration coefficient is 75% of the refrigeration coefficient of the Carnot reverse cycle at the same temperature limit. Try calculation:
    (1) heat release;
    (2) power consumption; and
    (3) the cooling capacity.

3. A Carnot heat pump provides 250 kW heat to the greenhouse in order to maintain the temperature of the chamber at 22 °C. Heat is taken from the outdoor air at 0 °C. Try to calculate the heating factor, the amount of energy consumed in the cycle, and the amount of heat absorbed from the outdoor air.

4. There is a reverse Carnot cycle, and the coefficient of performance (COP) is 4. Here are some questions: what is the ratio of the temperature of the high-temperature heat source to the low-temperature heat source? If the input power is 6 kW, what is the cooling capacity? If the system is circulating as a heat pump, what is the performance coefficient of the cycle and the amount of heat it can provide?

5. There are refrigerators that use the Braydon reverse cycle operating at temperatures between 300 and 250 K. If the cycle boost ratios are 3 and 6, try to calculate their COP. It is assumed that the working fluid can be regarded as an ideal gas, $c_p = 1.004$ kJ/(kg · K), $k = 1.4$.

Answers

1. Coefficient of refrigeration is

$$\varepsilon = \frac{T_2}{T_1 - T_2}$$

The temperature ratio of the high-temperature heat source to the low-temperature heat source is

$$\frac{T_1}{T_2} = \frac{6}{5}$$

The cooling capacity is

$$q_2 = \varepsilon w = 5 \times 6 = 30 \ \text{kW}$$

The heating coefficient is

$$\varepsilon' = \frac{q_1}{w_{net}} = \varepsilon + 1 = 6$$

The heating provided is

$$q_1 = \varepsilon' w_{net} = 6 \times 6 = 36 \ \text{kW}$$

2. (1) The refrigeration coefficient of Carnot reverse cycle with the same temperature limit is

$$\varepsilon_c = \frac{T_2}{T_1 - T_2} = \frac{245}{300 - 245} = 4.45$$

Then the cooling coefficient of the refrigerator is

$$\varepsilon = 75\%\varepsilon_c = 0.75 \times 4.45 = 3.34$$

The heat release is

$$\dot{Q}_1 = \dot{Q}_2 + \frac{\dot{Q}_2}{\varepsilon} = 9 + \frac{9}{3.34} = 11.69 \ \text{kW}$$

(2) The power consumption is

$$\dot{W} = \dot{Q}_1 - \dot{Q}_2 = 11.69 - 9 = 2.69 \, \text{kW}$$

(3) The cooling capacity is

$$\dot{Q}_2 = 9 \, \text{kW}$$

3. The heating coefficient is

$$\varepsilon' = \frac{T_1}{T_1 - T_2} = \frac{22 + 273}{22 - 0} = 13.4$$

The power consumption is

$$\dot{W} = \frac{\dot{Q}}{\varepsilon_c'} = \frac{250}{13.4} = 18.7 \ \text{kW}$$

The heat absorption is

$$\dot{Q}_2 = \dot{Q}_1 - \dot{W} = 250 - 18.7 = 231.3 \ \text{kW}$$

4. As we know,

$$(\text{COP})_R = \varepsilon = \frac{T_2}{T_1 - T_2} = 4$$

So, the ratio of the temperature of the two heat sources is

$$\frac{T_1}{T_2} = \frac{5}{4} = 1.25$$

The cooling capacity is

$$\dot{Q}_2 = \varepsilon \dot{W} = 4 \times 6 = 24 \, \text{kW}$$

When the system acts as a heat pump coefficient, the heating coefficient is

$$\varepsilon' = (\text{COP})_R = \frac{T_1}{T_1 - T_2} = \frac{1}{1 - \frac{T_2}{T_1}} = 5$$

The heat supply is

$$\dot{Q} = \varepsilon' \dot{W} = 5 \times 6 = 30 \; \text{kW}$$

5. Braydon reverse cycle cooling coefficient:

$$(\text{COP})_R = \frac{1}{\pi^{\frac{k-1}{k}} - 1}$$

When the boost ratio is 3,

$$(\text{COP})_R = \frac{1}{\pi^{\frac{k-1}{k}} - 1} = \frac{1}{3^{\frac{0.4}{1.4}} - 1} = 2.71$$

When the boost ratio is 6,

$$(\text{COP})_R = \frac{1}{\pi^{\frac{k-1}{k}} - 1} = \frac{1}{6^{0.4/1.4} - 1} = 1.50$$

Part 2: **Heat transfer**

Both *heat transfer* and *engineering thermodynamics* are sciences related to thermal phenomena. The difference between *heat transfer* and *engineering thermodynamics* exists in three aspects: the first is subject (the subjects of *engineering thermodynamics* are energy form, energy transformation, and the amount of heat transfer; the subjects of *heat transfer* are heat transfer rate, temperature variation/distribution, and cooling/heating time), the second is time (*engineering thermodynamics* gives no consideration of the time that a heat transfer process will take, but time is important in *heat transfer*), and the third is state (*engineering thermodynamics* cares about the equilibrium state, while *heat transfer* cares about the non-equilibrium state).

There are a variety of physical processes happening in the world we live. The most relevant to humans is heat transfer: for example, the formation of wind, frost, rain, and snow in nature is related to heat transfer, and typical thermal processes are carried out in boilers and blast furnaces in industrial production, all of which are closely related to the transfer process of heat. Heat transfer is the science studying the laws of heat flow, which is caused by temperature difference. The second law of thermodynamics points out that heat can be transferred spontaneously from a high-temperature heat source to a lower one. This means that there is heat transfer where there is a temperature difference.

In daily life, temperature differences are almost ubiquitous, so heat transfer is a universal phenomenon. It is necessary to discover the principles and characteristic technology behind heat transfer.

https://doi.org/10.1515/9783111329703-009

Chapter 9
Basic ways of heat transfer

Heat transfer is commonly encountered in daily life and engineering systems. For example, the greenhouse effect in nature and air conditioning in daily life, there are more fields involving heat transfer in industry, such as power, mechanical manufacture, chemical industry, refrigeration, architecture, environment, new energy source, microelectronics, nuclear energy, aerospace, microelectromechanical system, new material, military science and technology, life science, and biology technology. Energy can be transferred into two forms: work (W) and heat (Q). There are three basic ways of heat transfer: thermal conduction, thermal convection, and thermal radiation. In the actual heat transfer process, sometimes there is only one basic heat transfer way, for example, when we touch a cup containing hot water, we feel hot. This is because the heat of the hot water in the cup is transferred to the hands. In this process, only heat conduction works. Sometimes two or three basic ways are performed simultaneously. A good example of heat transfer in our daily life is shown in Fig. 9.1. It is clearly seen that there are three heat transfer ways at the same time: heat conduction is expressed through the touching action between hand and pot, heat convection occurs between the hot water near the bottom of the pot and the cold water in the upper part of the pot, and heat radiation from flame leads to the boiling evolution in the water.

Conduction   Convection

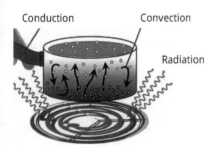

Radiation

Fig. 9.1: Heat transfer process in our daily life.

Heat exchangers are a typical application of heat transfer process, which are not only widely used in the industrial field but also can be commonly used in daily life (Fig. 9.2). A heat exchanger is an energy-saving device that realizes heat transfer between materials between two or more fluids at different temperatures. It can facilitate heat transfer from a fluid with a higher temperature to a fluid with a lower temperature so that the fluid temperature can meet the needs of process conditions, using heat exchangers can also improve the energy efficiency. The detailed introduction to heat exchangers will be elaborated in Chapter 13. A typical example of a heat exchanger is given next: the low temperature covers most of the time of winter in the

https://doi.org/10.1515/9783111329703-010

northern China, where the temperature can even reach minus tens of degrees. The indoor temperature can often be increased by a radiator. The radiator heating has an independent, sealed circulation system inside. The heated circulating water is connected to the radiator through pipes, and the radiator emits heat to the room at this time. After repeated thermal cycle work, the temperature of the entire room will rise evenly. Several heat transfer methods are included in the process of heating by heat exchanger. Think about what kinds of heat transfer ways are included in the heat exchanger?

Fig. 9.2: A photograph of commonly used heat exchanger.

Main contents in this chapter:
(1) The mechanism and characteristics of the three basic ways of heat transfer, heat convection, and thermal radiation.
(2) Basic concepts: heat flow rate, heat flux, thermal conduction, convection coefficient, surface heat transfer coefficient h, thermal conductivity k, and thermal resistance.
(3) Flexible use of three equations for the calculation of the relevant physical quantity: One-dimensional steady-state Fourier's law for plane wall, Newton's cooling law, and the heat flow rate through the plane wall.

9.1 Thermal conduction

9.1.1 Basic concepts in thermal conduction

Thermal conduction is the heat transfer phenomenon caused by the thermal movement of microparticles such as molecules, atoms, and free electrons inside or between the surfaces of objects in contact with each other. For example, if you hold one end of a metal rod and the other end is being heated by fire, the heat from fire will be transferred through the end you hold, causing you feel that the fingers are touching the fire (Fig. 9.3). This heat transfer way is named as heat conduction. Thermal conduction occurs either inside a solid or in a stationary liquid or gas.

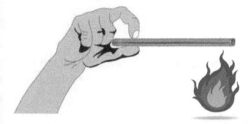

Fig. 9.3: Schematic diagram of thermal conduction.

The underlying mechanisms of heat conduction are different under different media. For gases and liquids, conduction is due to the collision and diffusion of the molecules during their random motion. For solids, conduction is due to the vibration of the molecules in a lattice (nonmetallic solids) and the energy transfer by free electrons (metal solids). The thermal conduction law under extremely small characteristic size is different from the macroscopic bulk, and here we only discuss the thermal conduction law under the macroscale.

In industry and daily life, the heat conduction of large flat walls is the simplest and the most common thermal problem. For example, heat is conducted through furnace walls in the steel industry, and in daily life, heat is conducted through the walls of houses. Here is an analysis example of one-dimensional steady-state thermal conductivity of large flat walls, as shown in Fig. 9.4.

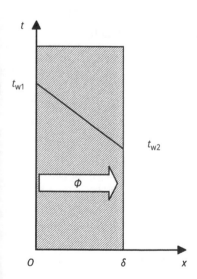

Fig. 9.4: Heat conduction of large flat walls.

Assuming:

(1) The temperature of the two surfaces of the flat wall is uniform and constant, which is denoted by t_{w1} and t_{w2}, respectively.

(2) The flat wall temperature only changes in the direction perpendicular to the wall surface.

(3) The flat wall temperature does not change over time.

(4) The heat transfer direction is the wall normal.

In order to quantify the magnitude of heat transfer, the heat transmitted per unit of time is called the heat flow rate Φ with unit of W. Experiments have shown that Φ can be represented by the following equation:

$$\Phi = A\lambda \frac{t_{w1} - t_{w2}}{\delta} \tag{9.1}$$

where A denotes the heat transfer area and λ is thermal conductivity of the material, which indicates the thermal conduction ability of the material. Different materials have different thermal conductivity values. The thermal conductivity of materials is related to many factors, and temperature is one of the main factors. In general, metal materials generally have the higher thermal conductivity, and a good conductive material (such as silver and copper) is also a good thermal conductive material; The thermal conductivity of liquid materials are slightly lower than metal materials, and gases have the smallest thermal conductivity. At room temperature, copper: 398 W/(m · K), wood: 0.35 W/(m · K).

As shown in eq. (9.1), the heat flow of the one-dimensional steady-state thermal conductivity of the plate is proportional to the surface area of the plate and the temperature difference between the two sides, inversely proportional to the thickness of the plate, and is related to the thermal conductivity of the plate material. The heat flux indicating heat flow per unit of time through unit area is defined by

$$q = \frac{\Phi}{A} = \lambda \frac{t_{w1} - t_{w2}}{\delta} \tag{9.2}$$

where q denotes heat flux with unit of W/m^2. The direction of heat flow is always from high temperature to low temperature under the second law of thermodynamics.

9.1.2 Definition of thermal resistance

Similar to electrical resistance, thermal resistance is an important concept in heat transfer. Thermal resistance of flat wall represents the object's resistance to heat conduction with unit of K/W. Equations (9.3) and (9.4) show the difference between thermal resistance and electrical resistance. Here are the calculation equations of thermal resistance and heat flux.

$$\text{Electrical resistance} = \frac{\text{potential difference}}{\text{current}} \qquad (9.3)$$

$$\text{Thermal resistance} = \frac{\text{temperature difference}}{\text{heat flux}} \qquad (9.4)$$

$$\text{Heat flux} = \frac{\text{temperature difference}}{\text{thermal resistance}} \qquad (9.5)$$

$$\Phi = A\lambda \frac{t_{w1} - t_{w2}}{\delta} = \frac{t_{w1} - t_{w2}}{\frac{\delta}{A\lambda}} = \frac{t_{w1} - t_{w2}}{R_\lambda} \qquad (9.6)$$

$$R_\lambda = \frac{\delta}{A\lambda} \qquad (9.7)$$

Thermal resistance analysis is an important tool to solve the problem of heat transfer. When solving the problem, it is often necessary to draw a thermal resistance network diagram. As an example of the steady-state thermal conductivity of the above-mentioned one-dimensional plate, the thermal resistance network of heat conduction is shown in Fig. 9.5.

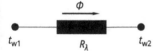

t_{w1} R_λ t_{w2} Fig. 9.5: Thermal resistance network for heat conduction.

9.2 Thermal convection

Free convection is the phenomenon of heat transfer caused by the macro-motion of fluid so that there is a relative displacement of the fluid under different temperatures. The free convection occurs only in the fluid and is accompanied by the thermal conduction of microscopic particles, such as boiling water in the kettle (Fig. 9.6).

Thermal convection is a heat transfer phenomenon occurred between viscous fluids and solid surfaces, which is the result of the combined effects of heat conduction and heat convection (Fig. 9.7). The difference between free convection and thermal convection is that free convection only occurs inside the fluid while the thermal convection occurs between fluid and solid surfaces. Thermal convection is a very interesting topic in industry.

According to the cause of the flow, convective heat transfer can be divided into two categories: natural convection and forced convection. Natural convection is caused by the different densities of each part of the fluid, such as cold air moving downwards and hot air moving upwards; Forced convection is the flow of fluids driven by fans, pumps, or other differential pressures, such as air flow caused by fan rotation, water in pipes flowing under pump drive, etc. In addition, the engineering also often needs to deal with the problem of liquid boiling on the hot surface or steam condensing on the cold

Fig. 9.6: A photo of free convection.

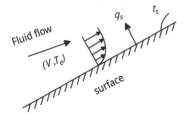

Fig. 9.7: Schematic diagram of thermal convection.

surface, which is called boiling heat transfer and condensation heat transfer, respectively. Phase transitions accompany these heat transfer problems.

9.2.1 Newton's cooling law

The basic calculation equation for convection heat exchange is Newton's cooling law (Fig. 9.8):

$$\Phi = Ah(t_w - t_f) \tag{9.8}$$

$$q = h(t_w - t_f) \tag{9.9}$$

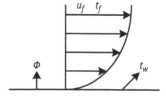

Fig. 9.8: Schematic diagram of Newton's cooling law.

where A is the contact area with unit of m², t_w is solid temperature with unit of °C, t_f is fluid temperature with unit of °C, and h is the surface heat transfer coefficient (usually called convection coefficients) with unite of W/(m² · K). It can reflect the strength of convective heat transfer. It is to be note that h is not a property of a material but depends on geometry, fluid velocity, fluid properties, etc. A major objective in the study of convection is to determine h.

Equation (9.8) only gives a definition of h, without revealing the various complex factors that affect h. The magnitude of h reflects the strength of the convective heat transfer and is related to the following factors:

(1) The properties of the fluid (thermal conductivity, viscosity, density, heat capacity, etc.)
(2) The form of fluid flow (laminar flow and turbulence flow)
(3) The cause of the flow (natural convection or forced convection)
(4) Shape and size of the surface of an object
(5) Whether there is phase-changed (boiling or condensing) during heat exchange process

In heat transfer, it is necessary to grasp the quantitative range of convective heat transfer coefficients in typical processes. How to determine the magnitude of h will be discussed in Chapter 11, and the h-range of some convective heat exchange processes is given in Tab. 9.1.

Tab. 9.1: The h-range of some convective heat exchange processes.

Type of convection	Convection coefficients h (W/(m² · K))
Air natural convection	1–10
Air forced convection	10–100
Water natural convection	100–1,000
Water forced convection	1,000–15,000
Water boiling	2,500–35,000
Water vapor condensation	5,000–25,000

As given in the table, the convective heat transfer of water is stronger than that of air, and the phase change is better than the non-phase change, and the forced convection is better than the natural convection.

9.2.2 Convection thermal resistance

In the previous section, we discussed the thermal resistance of the heat transfer process, and the expression of the thermal resistance can also be written during thermal convection. Newton's cooling law can be written in the form of Ohm's law expression as shown in eq. (9.10):

$$\Phi = \frac{t_w - t_f}{\frac{1}{Ah}}$$
(9.10)

$$R_h = \frac{1}{Ah}$$
(9.11)

where R_h is convection thermal resistance in K/W. For the case shown in Fig. 9.8, convection thermal resistance network is schematically described in Fig. 9.9.

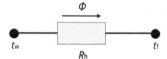

ϕ

t_w R_h t_f

Fig. 9.9: Convection thermal resistance network.

9.3 Thermal radiation

9.3.1 Basic concepts in thermal radiation

Energy is transmitted in the form of waves or movement of subatomic particles, Radiation is the phenomenon in which an object emits radiation energy outward by the stimulation of a factor. Radiation energy is emitted from the radiation source and radiates straight out in all directions. Thermal radiation is the phenomenon of emitted radiant energy from an object due to the thermal motion of microscopic particles inside an object.

There are two theories that explain radiation phenomena: electromagnetic theory and quantum theory. Electromagnetic theory points that radiation energy is the energy transmitted by electromagnetic waves; quantum theory points that radiation energy is the energy carried by discontinuous microscopic particles (photons).

Equation (9.12) is a mathematical description of electromagnetic waves:

$$c = \lambda v$$
(9.12)

where c is the speed of light in a medium, m/s; v is frequency, 1/s; λ is wavelength, μm.

The spectrum of electromagnetic waves is shown in Fig. 9.10.

Among them, visible light can be divided into seven colors of light (red, orange, yellow, green, cyan, blue, and purple) according to the wavelength, and infrared can be divided into near infrared ($0.76 < \lambda < 25$ μm) and far infrared ($25 < \lambda < 10^3$ μm). Microwave is a wave with a wavelength of $10^3 < \lambda < 10^6$ μm. Microwave ovens use microwaves to heat food, because microwaves can penetrate plastic, glass, and ceramic

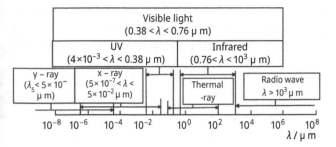

Fig. 9.10: The spectrum of electromagnetic waves.

products, but they will be absorbed by water molecules in the food to generate an internal heat source, which will evenly heat the food.

9.3.2 Definition of thermal radiation

In theory, the wavelength range of thermal radiation is 0 to ∞, but in the temperature range that is common in daily life and industry, the wavelengths of thermal radiation are mainly in the range between 0.1 and 100 μm, including three bands: part of ultraviolet, visible light, and part of infrared ray.

Thermal radiation is one of the basic ways for heat transfer, which differs a lot from heat conduction and thermal convection, it has the following characteristics:

(1) All objects with temperatures greater than 0 K (absolute zero) have the ability to emit thermal radiation, and the higher the temperature, the stronger the ability to emit thermal radiation. When releasing thermal radiation, the energy changes from internal thermal energy to radiant energy.

(2) All actual objects have the ability to absorb heat radiation. When absorbing thermal radiation, the energy changes from radiant energy to internal thermal energy.

(3) Objects need to be in contact with each other for heat conduction or thermal convection. Thermal radiation does not depend on intermediate media and can travel in a vacuum, and the transmission of radiant energy in a vacuum is most efficient. For example, the sun transmits heat to the earth in the form of thermal radiation.

(4) Heat transfer between objects by thermal radiation is bidirectional. When the temperature of two objects is different, the high-temperature object is emitting radiant energy to the low-temperature object, and the low-temperature object is also emitting radiation energy to the high-temperature object. Even when their temperatures are equal, the two objects are still constantly emitting and absorbing radiant energy, except that the heat transfer of the radiation is zero, which is a dynamic balancing process.

When there is a temperature difference between objects, the result of this energy exchange in the form of thermal radiation causes high-temperature objects to lose heat and low-temperature objects gain heat. This heat transfer phenomenon is called radiative heat transfer as shown in Fig. 9.11.

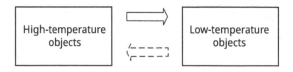

Fig. 9.11: Radiative heat transfer.

The main factor that affects the radiative heat transfer is the state of the object itself which include:
(1) Temperature and surface radiation characteristics of the object itself.
(2) Size, geometry, and relative position of an object.

9.4 Heat transfer process

We have introduced the three basic ways of heat transfer (thermal conduction, thermal convection, and thermal radiation). When actually analyzing the heat transfer process, the following three points have to be noted:
(1) The three basic modes of heat transfer: thermal conduction, thermal convection, and thermal radiation often do not occur separately. As mentioned earlier, convective heat transfer is the result of a combination of thermal conduction and thermal convection. Many heat transfer processes in real life and industrial production are combined by three heat transfer methods.
(2) When analyzing the heat transfer problem, figure out which heat transfer mode work, and calculate according to the rules of each heat transfer mode. The heat transfer process can be analyzed by using the thermal resistance network.
(3) If a certain heat transfer way has a very small effect compared to other heat transfer methods, it can often be ignored.

Heat transfer processes exist in various types of heat exchange equipment. In this section, the heat transfer process refers to the process by which heat is transferred from a fluid on one side of a solid wall to a fluid on the other side (Fig. 9.12).

This heat transfer process consists of three heat transfer modes connected in series:
(1) High-temperature fluid $\xrightarrow{\text{Convection}}$ High-temperature side of wall
(2) High-temperature side of wall $\xrightarrow{\text{Conduction}}$ Low-temperature side of wall
(3) Low-temperature side of wall $\xrightarrow{\text{Convection}}$ Low-temperature fluid

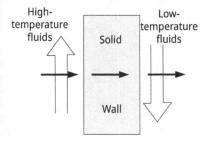

High-temperature fluids

Low-temperature fluids

Solid

Wall

Fig. 9.12: Heat transfer process from high-temperature fluids to low-temperature fluids.

9.4.1 The steady-state heat transfer process through the plane wall

Assume t_{f1}, t_{f2}, h_1, h_2 do not change with time, and λ is constant. The steady-state heat transfer process through the plane wall (Fig. 9.13) can be expressed as follows:

(1) Thermal convection on the left side:

$$\Phi = Ah_1(t_{f1} - t_{w1}) = \frac{t_{f1} - t_{w1}}{\frac{1}{Ah_1}} = \frac{t_{f1} - t_{w1}}{R_{h1}} \tag{9.13}$$

(2) Thermal conduction in the plane wall:

$$\Phi = A\lambda\frac{t_{w1} - t_{w2}}{\delta} = \frac{t_{w1} - t_{w2}}{\frac{\delta}{A\lambda}} = \frac{t_{w1} - t_{w2}}{R_\lambda} \tag{9.14}$$

(3) Thermal convection on the right side:

$$\Phi = Ah_2(t_{w2} - t_{f2}) = \frac{t_{w2} - t_{f2}}{\frac{1}{Ah_2}} = \frac{t_{w2} - t_{f2}}{R_{h2}} \tag{9.15}$$

Since it is a steady-state process, the heat flow Φ through each link of the series process is the same:

$$\Phi = \frac{t_{f1} - t_{f2}}{\frac{1}{Ah_1} + \frac{\delta}{A\lambda} + \frac{1}{Ah_2}} = \frac{t_{f1} - t_{f2}}{R_{h1} + R_\lambda + R_{h2}} = \frac{t_{f1} - t_{f2}}{R_k} \tag{9.16}$$

where $R_k = R_{h1} + R_\lambda + R_{h2}$, R_k is called overall thermal resistance.

The series thermal resistance network is similar to that of series electrical resistance network. The total thermal resistance is equal to the sum of each thermal resistance in series, as shown in Fig. 9.14.

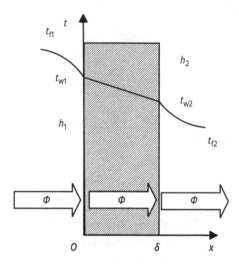

Fig. 9.13: Steady-state heat transfer process through the plane wall.

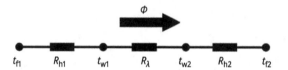

Fig. 9.14: Series thermal resistance network.

9.4.2 Heat transfer coefficient

Considering the whole heat transfer process, the calculation equation of heat transfer heat flow can be converted into:

$$\Phi = Ak(t_{f1} - t_{f2}) = Ak\Delta t \tag{9.17}$$

$$k = \frac{1}{\dfrac{1}{h_1} + \dfrac{\delta}{\lambda} + \dfrac{1}{h_2}} \tag{9.18}$$

where Δt is the temperature difference and k is called the overall heat transfer coefficient in W/(m² · K). The overall heat transfer coefficient is expressed as an indicator of the intensity of the heat transfer process, and it is numerically equal to the value of the heat flow when the temperature difference between the cold and heat flow is 1 °C and the heat transfer area is 1 m².

Using the overall heat transfer coefficient, the heat flux through the flat wall can be written as

$$q = k(t_{f1} - t_{f2}) = \frac{t_{f1} - t_{f2}}{\frac{1}{h_1} + \frac{\delta}{\lambda} + \frac{1}{h_2}} \tag{9.19}$$

Using the above equation, it is easy to obtain the heat flow through the plane wall Φ, heat flux q, and the wall temperatures t_{w1} and t_{w2}.

Here is an example: One-layer glass window, with 1.2 m, 1 m, and 0.3 mm of height, width and thickness, respectively. The thermal conductivity of glass is 1.05 W/(m · K); the air temperature indoors and outside are 20 °C and 5 °C, respectively; the convective heat transfer coefficients between indoor and outdoor air and glass windows are 5 W/(m^2 · K) and 20 W/(m^2 · K), respectively. Try to find the heat loss of the glass window, the thermal resistance of conductive heat transfer, and the thermal resistance of the convective heat transfer on both sides.

In this example, there are three heat transfer processes, namely heat convection between indoor hot air and the glass window, heat conduction along the glass window thickness direction, and heat convective between outdoor cold air and the glass window. Thermal resistance network is shown in Fig. 9.15.

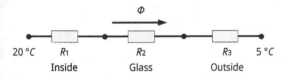

Fig. 9.15: Thermal resistance network of one-layer glass window.

The heat flow of the above system can be described by

$$\Phi = A \times \frac{t_{f_1} - t_{f_2}}{\frac{1}{h_1} + \frac{\delta}{\lambda} + \frac{1}{h_2}} 1.2 \times 1 \times \frac{20 - 5}{\frac{1}{5} + \frac{0.0003}{1.05} + \frac{1}{20}} = 71.9 \text{ W} \tag{9.20}$$

$$R_1 = \frac{1}{Ah_1} = \frac{1}{1.2 \times 5} = 0.167 \text{ K/W} \tag{9.21}$$

$$R_2 = \frac{\delta}{A\lambda} = \frac{0.0003}{1.2 \times 1.05} = 0.000238 \text{ K/W} \tag{9.22}$$

$$R_3 = \frac{1}{Ah_2} = \frac{1}{1.2 \times 20} = 0.0417 \text{ K/W} \tag{9.23}$$

where R_1, R_2, and R_3 denote the thermal resistance of conductive heat transfer, and that of the convective heat transfer on both sides.

9.5 Summary

(1) The mechanism and characteristics of the three basic ways of heat transfer, heat convection, and thermal radiation
(2) Basic concepts: heat flow rate, heat flux, thermal conduction, convection coefficient, surface heat transfer coefficient h, thermal conductivity λ, and thermal resistance
(3) Flexible use of three equations for the calculation of the relevant physical quantity: one-dimensional steady-state Fourier's law for plane wall, Newton's cooling law, and the heat flow rate through the plane wall

Exercises

1. When the shower head is working properly, the water supply is usually 1,000 cm³/min. Cold water is heated from 15 to 44 °C through the electric heater. What is the heating power of the electric heater calculated? To save energy, it has been proposed that the hot water after use (at 38 °C) be fed into a heat exchanger to heat the cold water going into the shower. If the heat exchanger can heat cold water to 27 °C, calculate how much energy can be saved by taking a bath for 15 min after using the waste heat recovery heat exchanger.

2. The surface area of a brick wall is 20 m², the thickness is 260 mm, and the average thermal conductivity is 1.5 W/(m · K). Set the surface temperature facing the room to be 25 °C and the outer surface temperature to be −10 °C, and calculate the heat flow from the brick wall to the outside world.

3. An isothermal integrated circuit chip with a length and width of 10 mm is mounted on a base plate, and air with a temperature of 25 °C cools the chip under the action of a fan. The maximum allowable temperature of the chip is 80 °C, and the surface heat transfer coefficient between the chip and the cooling air flow is 165 W/(m² · K). What is the maximum thermal design power of the chip without considering radiation? The thickness of the top surface of the chip is 1 mm higher than the bottom plate.

4. There is a gas cooler with a heat transfer coefficient h_1 = 95 W/(m² · K) on the air side surface, a wall thickness of 3 mm, λ = 47 W/(m · K), and a heat transfer coefficient h_2 = 5,750 W/(m² · K) on the water side surface. The wall can be seen as a flat wall, calculating the thermal resistance per unit area of each link and the total heat transfer coefficient from gas to water. Can you point out, in order to strengthen this heat transfer process, should start from which link?

5. In the previous problem, if a layer of ash with a thickness of 2 mm formed on the gas side, λ = 0.132 W/(m · K). A layer of scale with a thickness of 1 mm has formed on the water side, λ = 1.20 W/(m · K), and other conditions remain unchanged. Please calculate the total heat transfer coefficient at this time.

6. Consider a two-sided glass window, with 1.5 m, 1 m, and 0.3 mm of height, width, and thickness, respectively, the air interlayer is 5 mm thick, and the thermal con-

ductivity of air is 0.025 W/(m · K). The thermal conductivity of glass is 1.05 W/(m · K). The indoor and outside air temperatures are 30 °C and 5 °C, respectively. The convective heat transfer coefficients between indoor and outdoor air and glass windows are 10 W/(m² · K) and 25 W/(m² · K), respectively. Try to find the heat loss of the glass window and the heat conduction resistance of air interlayer.

7. The walls of a fruit storage room are made of corkboard and are 200 mm thick, with one wall measuring 3 m high and 6 m wide. In cold winter, the indoor temperature is set to 3 °C, the outdoor temperature is set to −15 °C, the surface heat transfer coefficient between the indoor wall and the environment is set to 6 W/(m² · K), and the surface heat transfer coefficient is set to 60 W/(m² · K) when there is a strong wind outside. Cork has a thermal conductivity of $\lambda = 0.044$ W/(m · K). Calculate the heat lost through this wall. The influence of outdoor wind weakening on wall heat dissipation is also discussed. (Hint: the outdoor surface heat transfer coefficient value can be taken as one half or one fourth of the original value for estimation.)

8. There is an ammonia evaporator with a heat transfer area of 20 m². The evaporation temperature of ammonia liquid is 0 °C, the inlet temperature of cooling water is 9.7 °C, the outlet temperature is 5 °C, and the heat transfer in the evaporator is 69,000 W. Please calculate the total heat transfer coefficient.

Answers

1. The heating power of the electric heater:

$$P = \frac{Q}{\tau} = \frac{cm\Delta t}{\tau} = \frac{4,180 \times 1,000 \times 1,000 \times 1,000 \times 10^{-6} \times (44-15)}{60}$$

$$= 2,019.6 \ W = 2.02 \ kW$$

The energy can be saved in 15 min:

$$Q = cm\Delta t = 4,180 \times 1,000 \times 1,000 \times 1,000 \times 10^{-6} \times 15 \times (27-15) = 752,400 \ J = 725.4 \ kJ$$

2. According to Fourier's law:

$$\Phi = \lambda A \times \frac{\Delta t}{\delta} 1.5 \times 20 \times \frac{25 - (-10)}{0.26} = 4,038.5 \ W$$

3. $\Phi = hA\Delta t = 165 \times [0.01 \times 0.01 + 4 \times (0.01 \times 0.001)] \times (80 - 25) = 1.2705 \ W$

4. Thermal resistance network diagrams are used to solve:

$$R_1 = \frac{1}{h_1} = \frac{1}{95} = 0.010526$$

$$R_2 = \frac{\delta}{\lambda} = \frac{0.003}{47} = 0.0000638$$

$$R_3 = \frac{1}{h_2} = \frac{1}{5,750} = 0.0001739$$

$$k = \frac{1}{\frac{1}{h_1} + \frac{\delta}{\lambda} + \frac{1}{h_2}} = 92.905 \ \mathrm{W}/(\mathrm{m}^2 \cdot \mathrm{K})$$

Heat transfer on the gas side surface should be enhanced.

5. $$k = \frac{1}{\frac{1}{h_1} + \frac{\delta_1}{\lambda_1} + \frac{\delta_2}{\lambda_2} + \frac{\delta_3}{\lambda_3} + \frac{1}{h_2}} = \frac{1}{\frac{1}{95} + \frac{0.002}{0.132} + \frac{0.003}{47} + \frac{0.001}{1.2} + \frac{1}{5,750}} = 37.385 \ \mathrm{W}/(\mathrm{m}^2 \cdot \mathrm{K})$$

It can be seen that scaling will reduce the heat transfer effect, so scaling should be avoided as far as possible in heat exchangers.

6. Thermal resistance network is shown in Fig. 9.16:

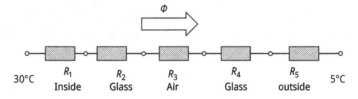

Fig. 9.16: Answer 6: additional graph.

$$Q = A \times \frac{t_{f_1} - t_{f_2}}{\frac{1}{h_1} + 2 \times \frac{\delta_1}{\lambda_1} + \frac{\delta_2}{\lambda_2} + \frac{1}{h_2}} = 1.5 \times 1 \times \frac{30 - 5}{\frac{1}{10} + 2 \times \frac{0.0003}{1.05} + \frac{0.005}{0.025} + \frac{1}{25}} = 85.17 \ \mathrm{W}$$

$$R_3 = \frac{\delta}{A \times \lambda} = \frac{0.005}{1.5 \times 0.025} = 0.133 \ \mathrm{K/W}$$

7. $$\Phi = \frac{\Delta t}{\frac{1}{h_1 A} + \frac{\delta}{A\lambda} + \frac{1}{A h_2}} = 68.505 \ \mathrm{W}$$

When the wind outside dies down, $h_2 = 30 \ \mathrm{W}/(\mathrm{m}^2 \cdot \mathrm{K})$

$$\Phi = \frac{\Delta t}{\frac{1}{h_1 A} + \frac{\delta}{A\lambda} + \frac{1}{A h_2}} = 68.28 \ \mathrm{W}$$

8. $$\overline{\Delta t} = \frac{\Delta t_1 + \Delta t_2}{2} = 7.35 \ ^\circ\mathrm{C}$$

According to $\Phi = kA\Delta t$

$$k = \frac{\Phi}{A\Delta t} = \frac{69,000}{20 \times 7.35} = 469.39 \ \mathrm{W}/(\mathrm{m}^2 \cdot \mathrm{K})$$

Chapter 10
Thermal conduction

Thermal conduction is the phenomenon of heat transfer due to the thermal motion of microscopic particles inside an object. From a microscopic point of view, the thermal conduction mechanisms for conductive solids, non-conductive solids, liquids, and gases are different.

In solids, the microscopic process of heat conduction can be described as the vibrational kinetic energy of the particles on the nodes in the crystal is larger under high temperature, while the kinetic energy of particle vibration is small under low temperature. Due to the interaction of particle vibrations, the thermal energy in the crystal is conducted from the particles with high kinetic energy to those with low kinetic energy. In conductors, due to the existence of a large number of free electrons, they are constantly maintaining random thermal motions. Generally, the energy of lattice vibration is small, and free electrons play a major role in the conduction of heat in metal crystals. Therefore, general electrical conductors are also good conductors for heat. The process of heat conduction in the liquid is as follows: the liquid molecules have strong thermal motion in the high temperature region. Due to the interaction between the liquid molecules, the energy of the thermal motion will gradually be transferred to the surrounding layers, causing the phenomenon of heat conduction. Due to the low thermal conductivity, it is similar to solids. The distance between gas molecules is relatively large, which is different from liquids. Gases rely on the random thermal motion of molecules and intermolecular collisions to transfer energy inside the gas, which finally forms macroscopic heat transfer.

Based on the hypothesis of continuous medium and from the macro point of view, the basic law and calculation method of heat conduction is discussed in this chapter. The continuous hypothesis is that the space occupied by real fluids or solids can be approximately regarded as continuous and filled with "particles" without voids. The macroscopic physical quantities (such as mass, velocity, pressure, and temperature) possessed by a particle satisfy all physical laws that should be followed, such as the law of conservation of mass and Newton's law of motion.

According to the continuous medium theory, particles are continuously filled with fluid or solids occupying the space. In general, most solids, liquids, and gases can be considered as continuous media. However, when the average free path of the molecules is not negligible compared with the macroscopic size of the object, such as rarefied gas with a reduced pressure to a certain degree, it cannot be considered a continuous medium, and thus it is not covered by this chapter.

https://doi.org/10.1515/9783111329703-011

10.1 The theoretical basis of thermal conduction

10.1.1 The basic law of thermal conductivity

10.1.1.1 Temperature field

Similar to velocity fields and gravitational fields, there is a temperature field in physics. In τ moment, the temperature distribution of all points within an object is called the temperature field of the object at that moment.

In general, a temperature field is a function of spatial coordinates and time. In a right-angle coordinate system, the temperature field can be expressed as

$$t = f(x, y, z, \tau) \tag{10.1}$$

Temperature field includes non-steady-state temperature field and steady-state temperature field. The temperature field in which temperature changes over time is called non-steady-state temperature field. For example, the temperature field occurs when the components of a thermal engine (internal combustion engine, steam turbine, aero engine, etc.) are started, stopped, or when the operating conditions are changed, and the thermal conduction of non-steady-state temperature field is called the non-steady-state heat conduction. The temperature field in which temperature does not change over time is called steady-state temperature field, and the thermal conduction of steady-state temperature field is called steady-state heat conduction. Steady-state temperature field

$$t = f(x, y, z) \tag{10.2}$$

$$\frac{\partial t}{\partial \tau} = 0 \tag{10.3}$$

According to the change of temperature in three directions of space, the temperature field can be classified into one-dimensional temperature field, two-dimensional temperature field, and three-dimensional temperature field. The differences between non-steady state and steady state are shown in Tab. 10.1.

Tab. 10.1: The differences between non-steady state and steady state.

	Non-steady state	Steady state
One-dimensional temperature field	$t = f(x, \tau)$	$t = f(x)$
Two-dimensional temperature field	$t = f(x, y, \tau)$	$t = f(x, y)$
Three-dimensional temperature field	$t = f(x, y, z, \tau)$	$t = f(x, y, z)$

10.1.1.2 Isothermal surface and isothermal line

At the same moment, a line or face connected by the same point of temperature in the temperature field is called an isothermal line or an isothermal surface. It has the following characteristics:

(a) Any line on the isothermal surface is isothermal, and the isothermal surface behaves as isothermal line on any two-dimensional cross-section. If a plane and a set of isothermal surfaces intersect, a set of isothermal lines with different temperatures is obtained.

(b) The temperature field can be represented by a set of isothermal surfaces or isothermal lines as shown in Fig. 10.1.

(c) At the same time, the isothermal surfaces or isothermal lines of different temperatures in the object cannot intersect because it is impossible for any point to have two or more temperature values simultaneously.

(d) Under the continuous medium hypothesis, the isothermal surface (or isothermal line) in an object either forms a closed surface (curve) or terminates at the object's boundary.

(e) When the temperature intervals of each two adjacent isothermal surfaces (isothermal lines) are equal in the figure, the size of heat flows in different regions can be reflected by the density of the isothermal surfaces (isothermal lines).

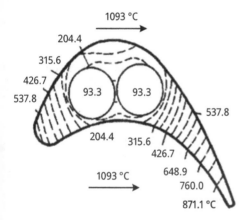

Fig. 10.1: Isothermal line.

10.1.1.3 Temperature gradient

As shown in Fig. 10.2, in the temperature field, the temperature change rate in a certain direction x can be expressed by the partial derivative:

$$\frac{\partial t}{\partial x} = \lim_{x \to 0}\left(\frac{\Delta t}{\Delta x}\right)$$

(10.4)

Obviously, the temperature change rate is the largest and the temperature change is the most severe in the normal direction of the isothermal surface, and thus temperature change rate vector in the normal direction of the isotherms is called temperature gradient:

$$\text{grad } t = \frac{\partial t}{\partial n} \boldsymbol{n} \tag{10.5}$$

where $\boldsymbol{n}$ is the unit vector in the normal direction of isotherms, which is the direction of the rapidest increase direction of the temperature. The temperature gradient is a vector that points to an increase in temperature as shown in Fig. 10.2.

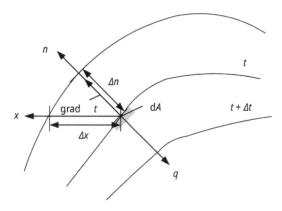

Fig. 10.2: Temperature gradient.

In a right-angle coordinate system, the temperature gradient can be expressed as

$$\text{grad } t = \frac{\partial t}{\partial x} \boldsymbol{i} + \frac{\partial t}{\partial y} \boldsymbol{j} + \frac{\partial t}{\partial z} \boldsymbol{k} \tag{10.6}$$

The three terms on the right of the above formula are the partial derivatives in x, y, and z directions, respectively; $\boldsymbol{i}$, $\boldsymbol{j}$, and $\boldsymbol{k}$ are unit vectors in the x, y, and z directions, respectively.

10.1.1.4 Heat flux

As shown in Fig. 10.3, dA denotes the area of a micro-element on the isothermal surface t, the heat flow rate vertically through dA is $d\Phi$, its direction points in the direction of the decrease in temperature. The thermal conductive heat flux on dA can be expressed as

$$q = \frac{d\Phi}{dA} \tag{10.7}$$

The magnitude and direction of the heat flux can be expressed in the heat flux vector $\boldsymbol{q}$:

$$q = -\frac{d\Phi}{dA}n \qquad (10.8)$$

The direction of the heat flux vector points to the direction of temperature reduction as shown in Fig. 10.3.

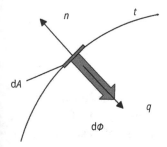

Fig. 10.3: The direction of heat flux vector.

In a right-angle coordinate system, the heat flux vector can be expressed as

$$q = q_x i + q_y j + q_z k \qquad (10.9)$$

where q_x, q_y, q_z represent the magnitude of the components of q in the three coordinate directions.

10.1.2 The basic law of thermal conductivity

On the basis of a large number of experimental studies on the thermal conduction process, a French scientist named Fourier proposed the famous fundamental law of heat conduction Fourier's law in 1822 and pointed out the relationship between the heat flux vector and temperature gradient. For isotropic objects whose physical property parameters do not change with direction, Fourier's law is expressed as

$$q = -\lambda \operatorname{grad} t = -\lambda \frac{\partial t}{\partial n} n \qquad (10.10)$$

Fourier's law shows that the magnitude of heat flux is proportional to the absolute value of the temperature gradient, and its direction is opposite to the direction of the temperature gradient. The scale factor is the thermal conductivity. Fourier's law in scalar form can be expressed as

$$q = -\lambda \frac{\partial t}{\partial n} \qquad (10.11)$$

For isotropic materials, the thermal conductivity λ in all directions is the same, and thus the equations above can be deformed to the following equations:

$$q = -\lambda \operatorname{grad} t \atop \operatorname{grad} t = \frac{\partial t}{\partial x} i + \frac{\partial t}{\partial y} j + \frac{\partial t}{\partial z} k \Big\} \Rightarrow \begin{cases} q = -\lambda \left(\frac{\partial t}{\partial x} i + \frac{\partial t}{\partial y} j + \frac{\partial t}{\partial z} k \right) \\ q = q_x i + q_y j + q_z k \end{cases}$$

$$q_x = -\lambda \frac{\partial t}{\partial x} \tag{10.12}$$

$$q_y = -\lambda \frac{\partial t}{\partial y} \tag{10.13}$$

$$q_z = -\lambda \frac{\partial t}{\partial z} \tag{10.14}$$

According to Fourier's law, it should be defined clearly for the thermal conductivity of the material and the temperature field in order to calculate the heat flux or heat flow. Therefore, solving the temperature field is the main task of heat conduction analysis.

Think about how to derive a one-dimensional steady-state thermal formula:

$$q = \lambda \frac{t_{w1} - t_{w2}}{\delta}$$

Terms of application of Fourier's law:
(1) Fourier's law applies only to isotropic objects (Fig. 10.4). In anisotropic objects, since the thermal conductivity may vary with the direction, the direction of the heat flux vector is not only related to the temperature gradient but also to the directionality of the thermal conductivity. The analysis of this type of thermal conduction problem is more complicated;
(2) Fourier's law is applicable to general steady-state and non-steady-state heat conduction problems in engineering technology. For thermal conduction problems at extremely low temperatures and transient heat conduction processes that produce extremely high heat flux density in a short time, such as high power and short pulses (pulse width up to $10^{-12} - 10^{-15}$ s) laser transient heating, Fourier's law no longer applies.

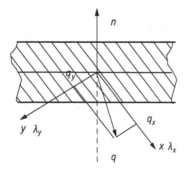

Fig. 10.4: According to Fourier's law, heat flow is decomposed.

10.1.3 Thermal conductivity

Thermal conductivity is one of the important thermophysical parameters of substances, which is the magnitude of the material's thermal conduction ability. According to Fourier's law:

$$\lambda = \frac{q}{|\text{grad } t|} \tag{10.15}$$

The thermal conductivity values of most materials can be measured experimentally. There are two types of methods for measuring thermal conductivity: steady-state method and non-steady-state method. Fourier's law of thermal conduction is the basis of the steady-state method. The values of thermal conductivity of various materials vary greatly. Table 10.2 summarizes the thermal conductivity values of some typical materials at room temperature.

Tab. 10.2: Thermal conductivity values of several typical materials at 20 °C.

Material name	λ [W/(m · K)]	Material name	λ [W/(m · K)]
Metal (solid):		Pine (parallel wood grain)	0.35
Sterling silver	427	Ice (0 °C)	0.22
Pure copper	398	**Liquid:**	
Brass (70% Cu, 30% Zn)	109	Water (0 °C)	0.551
Pure aluminum	236	Mercury	7.90
Aluminum alloy (87% Al, 13% Si)	162	Transformer oil	0.124
Pure iron	81.1	Diesel oil	0.128
Carbon steel (about 0.5% C)	49.8	Lubricating oil	0.146
Non-metal (solid):		**Gas (atmospheric pressure):**	
Quartz crystal (0 °C, parallel to the axis)	19.4	Air	0.0257
Quartz glass (0 °C)	1.13	Nitrogen	0.0256
Marble	2.70	Hydrogen	0.177
Glass	0.65–0.71	Water vapor (0 °C)	0.183
Pine (vertical grain)	0.15		

According to Tab. 10.2, the thermal conductivity of a substance has the following characteristics numerically:

(1) For the same substance, the thermal conductivity value of the solid state is the largest and the thermal conductivity value of the gaseous state is the smallest. For example, also at 0 °C, the thermal conductivity of ice is the highest, followed by the thermal conductivity of water, and the thermal conductivity of water vapor is the lowest.

(2) Thermal conductivity of general metals > thermal conductivity of non-metallic materials. This is because the heat conduction mechanism of metals and non-metallic materials is different. The heat conduction of metals mainly depends on

the movement of free electrons, while the heat conduction of non-metallic materials mainly depends on the vibration of molecules or lattices.

(3) A metal with good conductive properties and its thermal conductivity is good.
(4) Thermal conductivity of pure metals > the thermal conductivity of its alloys. This is due to impurities (or other metals) in the alloy that disrupts the structure of the lattice or hinders the movement of electrons.
(5) For anisotropy objects, the value of thermal conductivity is related to the direction.
(6) For the same substance, the thermal conductivity of the crystal > thermal conductivity of amorphous objects.

There are many factors that affect thermal conductivity, such as substance types, structures, and physical states. Since heat conduction is carried out in a non-uniform temperature field, the effect of temperature on thermal conductivity is particularly important. In the common temperature range of industry and daily life, the thermal conductivity of most materials can be approximated as a linear relationship with temperature (Fig. 10.5), which is expressed as

$$\lambda = \lambda_0 [1 + bt] \tag{10.16}$$

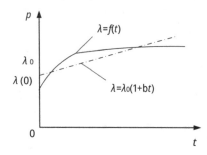

Fig. 10.5: Relationship between thermal conductivity and temperature.

where λ_0 is a calculated value under 0 °C as above formula; b is a constant determined by an experiment, the value of which is related to the type of substance.

Most building materials and thermal insulation materials (or heat insulation materials) have porous or fibrous structures (such as brick, concrete, asbestos, and slag) and are not homogeneous media, collectively referred to porous materials. Porous materials have the advantages of light weight, thermal insulation, low price, and easy to construct. In recent years, China has also produced various new thermal insulation materials (such as rock wool boards, expanded plastics, and aerogel). Strictly speaking, the porous structure of the material is no longer a homogeneous continuous medium, and the thermal conductivity of a porous material refers to its apparent thermal conductivity. At high temperatures, the heat conduction occurs through solid skeleton structure and through air contained in pores. Besides heat conduction, thermal radiation also exists for porous materials. Materials used for insulation are called insulation materials. The regulations on the thermal conductivity of insulation materials are different in various countries in the world, and Chinese national standards that stipulate those materials with a thermal conductivity of less than 0.12 W/(m · K) at temperatures below 350 °C are called insulation materials.

10.1.4 Mathematical description of thermal conduction problems

A complete mathematical description of a specific heat transfer process (i.e., a mathematical model of heat transfer) should include thermal conduction differential equation and single-value condition. Thermal conduction differential equation is a general equation that the temperature field of all thermally conductive objects should satisfy. For specific issues, corresponding time and boundary conditions must also be specified.

Building a reasonable mathematical model is the first and most important step to solve the thermal problem. The mathematical model is solved to obtain the temperature field of the object, and the corresponding heat flux distribution can be determined according to Fourier's law.

At present, the most widely used methods for solving heat conduction problems include:
(1) Analytical solutions
(2) Numerical solutions
(3) Experimental methods

The analytical solution is calculated according to the function, and the value of the dependent variable can be found for a given argument. Numerical solutions can be used in problems that are difficult to express in mathematical expressions or that formulas are difficult to calculate. The solution to a numerical solution is to guess the solution first and then test whether the solution is sufficient to solve the problem. This chapter mainly introduces the analytical solution.

In order to obtain a mathematical expression of the temperature field of a thermally conductive object, the variation relation of the temperature field of an object must be established according to the law of conservation of energy and Fourier's law. This expression is called the thermal conduction differential equation. The derivation of the differential equation is introduced next.

Assume:
(1) The object consists of an isotropic continuous medium
(2) With internal heat source with strength Φ, representing the generated heat in unit time and unit volume, W/m^3

As shown in Fig. 10.6, in the right-angle coordinate system, a parallel hexagonal micro-element is selected as the research object, and its edge lengths are dx, dy, and dz, respectively. $d\Phi_x$, $d\Phi_y$, and $d\Phi_z$ are the heat of the micro-element introduced in the x, y, and z directions per unit time, and $d\Phi_{x+dx}$, $d\Phi_{y+dy}$, and $d\Phi_{z+dz}$ are the heat exported from the micro-element in the x, y, and z directions, respectively, in the unit time.

In unit time, the sum of net heat flow imported to micro-element $d\Phi_\lambda$ and heat generation $d\Phi_V$ is equal to the increase in elemental thermodynamic energy dU. The thermal balance of micro-elements during thermal conduction can be expressed as

$$d\Phi_\lambda + d\Phi_V = dU \tag{10.17}$$

$$d\Phi_\lambda = d\Phi_{\lambda x} + d\Phi_{\lambda y} + d\Phi_{\lambda z} \tag{10.18}$$

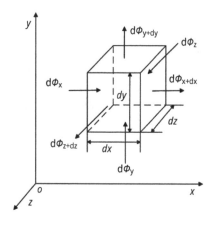

Fig. 10.6: Schematic diagram of a small volume with heat flux from three directions.

Expression of thermal equilibrium based on micro-element $d\Phi_\lambda + d\Phi_V = dU$

$$\rho c \frac{\partial t}{\partial \tau} = \left[\frac{\partial}{\partial x}\left(\lambda \frac{\partial t}{\partial x}\right) + \frac{\partial}{\partial y}\left(\lambda \frac{\partial t}{\partial y}\right) + \frac{\partial}{\partial z}\left(\lambda \frac{\partial t}{\partial z}\right) \right] + \dot{\Phi} \tag{10.19}$$

The differential equation of heat conduction establishes the function relationship of the temperature of the object with time and space during the heat conduction process.

When the thermal conductivity λ is constant, thermal differential equation can be simplified to

$$\frac{\partial t}{\partial \tau} = \frac{\lambda}{\rho c}\left(\frac{\partial^2 t}{\partial x^2} + \frac{\partial^2 t}{\partial y^2} + \frac{\partial^2 t}{\partial z^2}\right) + \frac{\dot{\Phi}}{\rho c} \tag{10.20}$$

or

$$\frac{\partial t}{\partial \tau} = a\nabla^2 t + \frac{\dot{\Phi}}{\rho c} \tag{10.21}$$

where ∇^2 is Laplace operator. In a right-angle coordinate system:

$$\nabla^2 t = \frac{\partial^2 t}{\partial x^2} + \frac{\partial^2 t}{\partial y^2} + \frac{\partial^2 t}{\partial z^2} \qquad (10.22)$$

a is thermal diffusivity, m²/s:

$$a = \frac{\lambda}{\rho c} \qquad (10.23)$$

Thermal diffusivity is an important thermophysical parameter in the process of non-steady heat conduction. Its size reflects how fast the temperature changes when the object is heated or cooled transiently. The greater the thermal diffusivity, the faster the temperature change. For example, if both hands hold the same length and thickness of the copper rod and the wooden stick, and the other end is simultaneously extended into the hot flame, the hand holding the copper rod will feel the heat first (copper: 1.5×10^{-7} m²/s, wood: 5.33×10^{-5} m²/s).

For special cases, the thermal conduction differential equations (eq. (10.20)) can be further simplified.

(1) Object has no internal heat source, $\dot{\Phi} = 0$

$$\frac{\partial t}{\partial \tau} = a\nabla^2 t \qquad (10.24)$$

(2) Steady-state heat conduction, $\dfrac{\partial t}{\partial \tau} = 0$

$$a\nabla^2 t + \frac{\dot{\Phi}}{\rho c} = 0 \qquad (10.25)$$

(3) Steady-state heat conduction, no internal heat source

$$\frac{\partial^2 t}{\partial x^2} + \frac{\partial^2 t}{\partial y^2} + \frac{\partial^2 t}{\partial z^2} = 0 \qquad (10.26)$$

10.2 Steady-state heat conduction

Steady-state heat conduction refers to the heat conduction process in which the temperature field does not change with time.

The main content of this section is to discuss the one-dimensional steady-state heat conduction problems of plane walls, cylindrical walls, and fins that are common in daily life and engineering.

When the two surfaces of the plane wall, respectively, maintain a uniform and constant temperature, the heat conduction of the flat wall is one-dimensional steady-state heat conduction that includes:

(1) Single-layer plane wall
(2) Multilayer plane wall

The following is an analysis of the one-dimensional steady-state thermal conduction problems of single-layer and multilayer flat walls, respectively.

10.2.1 Steady heat conduction of plane wall

(1) Steady-state heat conduction of single-layer plane wall (Fig. 10.7):
Assume: Surface area A, thickness δ, λ is constant, no internal source, and the surface on both sides maintains a uniform and constant temperature t_{w1}, t_{w2}, and $t_{w1} > t_{w2}$.
   Select the coordinate axis x perpendicular to the wall, as shown in Fig. 10.7.

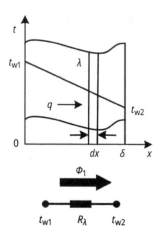

Fig. 10.7: Heat conduction through single-layer plane wall.

Build governing equations and write out boundary conditions:

$$\frac{\partial^2 t}{\partial x^2} + \frac{\partial^2 t}{\partial y^2} + \frac{\partial^2 t}{\partial z^2} = 0$$

$$\begin{cases} \frac{d^2 t}{dx^2} = 0 \\ x = 0, t = t_{w1} \\ x = \delta, t = t_{w2} \end{cases}$$

Using the above equations, the solution of the temperature distribution within the flat wall can be obtained:

$$t = t_{w1} - \frac{t_{w1} - t_{w2}}{\delta} x \qquad (10.27)$$

When λ is constant, the temperature distribution curve in the flat wall is a straight line with a slope

$$\frac{dt}{dx} = -\frac{t_{w1} - t_{w2}}{\delta} \tag{10.28}$$

The heat flux obtained by Fourier's Law is

$$q = -\lambda \frac{dt}{dx} = \lambda \frac{t_{w1} - t_{w2}}{\delta} \tag{10.29}$$

The heat flow through the entire flat wall is

$$\Phi = Aq = A\lambda \frac{t_{w1} - t_{w2}}{\delta} \tag{10.30}$$

(2) Steady-state heat conduction of multilayer plane wall (Fig. 10.8):
The multilayer flat wall is composed of multiple layers of different materials. For example, the furnace wall of the boiler is generally divided into a refractory layer, an insulation layer, and an ordinary brick exterior wall, and for the large boiler, a layer of steel plate is also wrapped outside. When the two surfaces are maintained at a uniform and constant temperature, the heat conduction is also one-dimensional steady-state heat conduction.

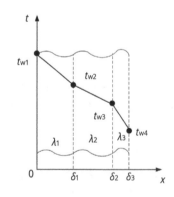

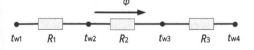

Fig. 10.8: Heat conduction through three-layer plane wall.

Take the three-layer plane wall into account, for example, assume:
(1) Thickness of each layer are δ_1, δ_2, δ_3, thermal conductivity of each layer of material is λ_1, λ_2, λ_3, and all are constant;

(2) The contact between the layers is close, and the surface of each other's contact has the same temperature;
(3) The outer surfaces on both sides of the flat wall maintain a uniform and constant temperature t_{w1}, t_{w4}.

Obviously, the heat conduction through the three-layer flat wall is steady-state heat conduction, and the heat flux of each layer is the same.

Total thermal resistance is the sum of thermal resistance of each layer. It can be represented by the thermal resistance network diagram, as shown in Fig. 10.8.

From the calculation formula of single-layer plane-wall steady-state heat conduction:

$$\Phi = \frac{t_{w1} - t_{w4}}{R_{\lambda 1} + R_{\lambda 2} + R_{\lambda 3}} = \frac{t_{w1} - t_{w4}}{\frac{\delta_1}{A\lambda_1} + \frac{\delta_2}{A\lambda_2} + \frac{\delta_3}{A\lambda_3}} \tag{10.31}$$

By analogy, for the steady-state heat conduction of n-layer flat walls:

$$\Phi = \frac{t_{w1} - t_{w(n+1)}}{\sum_{i=1}^{n} R_{\lambda i}} \tag{10.32}$$

Using the concept of thermal resistance, it is easy to find the heat flux through the steady-state heat conduction of multiple flat walls and then the temperature of the interface between the layers.

10.2.2 Steady-state heat conduction of the cylinder wall

Circular pipes are widely used in industry and daily life. This part mainly discusses the temperature distribution in the wall and the heat flow during the steady-state heat conduction of the cylinder wall.

(1) Steady-state heat conduction of a single-layer cylindrical wall (Fig. 10.9):
Assume: Inner and outer radii are r_1, r_2, length is l, λ is constant, no internal heat source, inner and outer wall temperature t_{w1}, t_{w2} are uniform and constant.

In cylindrical coordinate, the temperature inside the wall changes only along the radial direction, the heat conduction in the cylinder wall is one-dimensional steady-state heat conduction.

Build mathematical models and write out boundary conditions:

$$\begin{cases} \frac{d}{dr}\left(r\frac{dt}{dr}\right) = 0 \\ r = r_1, t = t_{w1} \\ r = r_2, t = t_{w2} \end{cases}$$

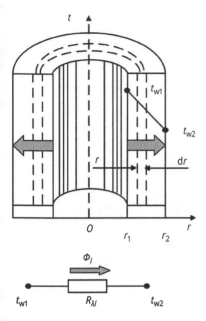

Fig. 10.9: Heat conduction through single-layer cylindrical wall.

Integrating the thermal differential equation twice, the general solution is

$$t = C_1 \ln r + C_2 \qquad (10.33)$$

The temperature inside the cylinder wall is distributed as a logarithmic curve. Substituting the boundary conditions, the temperature distribution can be expressed as

$$t = t_{w1} - (t_{w1} - t_{w2}) \frac{\ln\left(\frac{r}{r_1}\right)}{\ln\left(\frac{r_2}{r_1}\right)} \qquad (10.34)$$

The rate of temperature change in the direction of r is

$$\frac{dt}{dr} = -\frac{t_{w1} - t_{w2}}{\ln\left(\frac{r_2}{r_1}\right)} \frac{1}{r} \qquad (10.35)$$

Its absolute value is gradually decreasing in the direction of r.

According to Fourier's law, the heat flux through the cylinder wall r is

$$q = -\lambda \frac{dt}{dr} = \lambda \frac{t_{w1} - t_{w2}}{\ln\left(\frac{r_2}{r_1}\right)} \frac{1}{r} \qquad (10.36)$$

From the above equation, it is noted that the radial heat flux is a function of r, and as r increases, the heat flux gradually decreases. But for steady-state thermal conduction, the heat flow through the entire cylinder wall is constant:

$$\Phi = 2\pi rlq = \frac{t_{w1} - t_{w2}}{\frac{1}{2\pi\lambda l}\ln\frac{r_2}{r_1}} = \frac{t_{w1} - t_{w2}}{\frac{1}{2\pi\lambda l}\ln\frac{d_2}{d_1}} = \frac{t_{w1} - t_{w2}}{R_\lambda} \tag{10.37}$$

R_λ is the total thermal resistance of the entire cylinder wall, K/W.
The heat flow of the unit-length cylinder wall is

$$\Phi_l = \frac{\Phi}{l} = \frac{t_{w1} - t_{w2}}{\frac{1}{2\pi\lambda}\ln\frac{d_2}{d_1}} = \frac{t_{w1} - t_{w2}}{R_{\lambda l}} \tag{10.38}$$

$R_{\lambda l}$ is the thermal resistance of the wall of the unit length of the cylinder, K/(W · m).

(2) Steady-state heat conduction of a multilayer cylindrical wall:
Using the concept of thermal resistance, it is easy to analyze the problem of steady-state thermal conductivity.

Take a three-layer cylindrical wall as an example, as shown in Fig. 10.10. There is no internal heat source, and the thermal conductivity of each layer λ_1, λ_2, λ_3 is constant, and the inner and outer wall surfaces maintain a uniform and constant temperature t_{w1}, t_{w4}. This is obviously also a one-dimensional steady-state heat conduction problem. The heat flow through the cylinder walls of each layer is equal, and the total thermal resistance is equal to the sum of the thermal resistances of the layers. It can be described by the thermal resistance network diagram, as shown in Fig. 10.10. The heat flow of the unit-length cylinder wall is

$$\Phi_l = \frac{t_{w1} - t_{w4}}{R_{\lambda l_1} + R_{\lambda l_2} + R_{\lambda l_3}} = \frac{t_{w1} - t_{w4}}{\frac{1}{2\pi\lambda_1}\ln\frac{d_2}{d_1} + \frac{1}{2\pi\lambda_2}\ln\frac{d_3}{d_2} + \frac{1}{2\pi\lambda_3}\ln\frac{d_4}{d_3}} \tag{10.39}$$

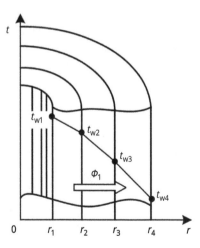

Fig. 10.10: Heat conduction through multilayer cylindrical wall.

10.3 Summary

(1) Fourier's law expression and its applicable conditions
(2) Numerical range and characteristics of thermal conductivity of objects
(3) Mathematical description (mathematical model) of heat conduction problem
(4) Analysis and solution of one-dimensional steady-state heat conduction problems of flat walls and cylindrical walls

Exercises

1. Fig. 10.11 is a diagram of the device that measures the thermal conductivity of the material by comparison. The thickness of the standard test sample $\delta_1 = 15$ mm, thermal conductivity $\lambda_1 = 0.15$ W/(m · K). The thickness of the sample to be tested $\delta_2 = 16$ mm. The edge of the specimen is extremely hot. Wall temperature measured in steady-state $t_{w1} = 45$ °C, $t_{w2} = 23$ °C, and $t_{w3} = 18$ °C. Ignore the cooling loss at the edge of the specimen. Try the thermal conductivity of the parts to be tested.

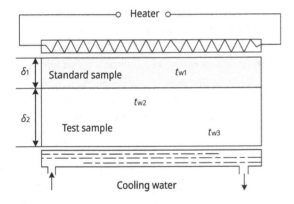

Fig. 10.11: Exercise 1: additional graph.

2. There is a flat wall with a thickness of 20 mm, which has a thermal conductivity of 1.3 W/(m · K). In order to make the heat loss per square meter wall not exceed 1,200 W, a layer of thermal insulation with thermal conductivity of 0.12 W/(m · K) is covered on the outer surface. Temperatures on both sides of the composite wall are known to be 656 and 100 °C, respectively. Calculate the total thickness range of the covered insulation layer.

3. Boil the water in a pan with a bottom temperature of 111 °C and a heat flow density of 45,200 W/m² in contact with the water. After using it for a period of time, a layer of scale with an average thickness of 3 mm is formed at the bottom of the pot, assuming that the surface temperature and heat flow density of the scale in

contact with the water at this time are equal to the original values, respectively, calculate the temperature of the contact surface between the scale and the metal pot bottom. Assuming the thermal conductivity of scale is 1.5 W/(m · K).

4. The double-glazed window system consists of two layers of glass with a thickness of 6 mm and the air gap between them, and the air gap thickness is 8 mm. Suppose that the indoor-facing glass surface temperature and the outdoor glass surface temperature are 20 and –10 °C, respectively. Try to determine the heat loss of the double-glazed window. If a single-layer glass window is used, other conditions remain unchanged, and how many times the heat loss is double-layered glass? The glass window size is 70 cm × 60 cm. Natural convection in the air gap is not taken into account. The thermal conductivity of the glass is 0.78 W/(m · K), and the thermal conductivity of the air is 0.023 W/(m · K).

5. A copper wire with a diameter of 3 mm has a resistance of 0.003 Ω/m long. The wire is surrounded by an insulating layer with a thickness of 1 mm and a thermal conductivity of 0.15 W/(m · K). Assume that the insulation can withstand a maximum temperature of 65 °C and a minimum temperature of 0 °C. Calculate the maximum current allowed through the wire under this condition.

6. The thermal power plant has a superheated steam pipe (steel pipe) with an outer diameter of 100 mm, which is covered with mineral wool of the thermal conductivity $\lambda = 0.05$ W/(m · K) to ensure the temperature. The temperature of the outer surface of the pipe is known to be 500 °C. Try to confirm the thickness of the mineral wool under the request that the temperature of mineral wool outside surface is under 50 °C, and the heat loss of per meter pipe is under 180 W.

7. In addition to a thermal pipe with an outer diameter of 100 mm, the thermal conductivity of one material covering two layers of insulation is 0.06 W/(m · K), the thermal conductivity of the other material is 0.12 W/(m · K), and the thickness of both materials is taken as 75 mm. Compare the effects of the two methods of attaching materials with small thermal conductivity to the pipe wall and materials with large thermal conductivity to the pipe wall. Does this effect exist for flat walls? Suppose that the total temperature difference between the inner and outer surfaces of the insulation remains constant in both approaches.

Answers

1. $q = \dfrac{t_{w_1} - t_{w_2}}{\frac{\delta_1}{\lambda_1}} = \dfrac{t_{w_2} - t_{w_3}}{\frac{\delta_2}{\lambda_2}} \Rightarrow \dfrac{45 - 23}{\frac{0.015}{0.15}} = \dfrac{23 - 18}{\frac{0.016}{\lambda_2}}$

$\lambda_2 = 0.704$ W/m · K

2. $q = \dfrac{t_1 - t_2}{\frac{\delta_1}{\lambda_1} + \frac{\delta_2}{\lambda_2}} = \dfrac{656 - 100}{\frac{0.02}{1.3} + \frac{\delta_2}{0.12}} \leq 1,200 \Rightarrow \delta_2 \geq 0.05375$ m

3. $q = \dfrac{t_w - 111}{\frac{0.003}{1.5}} = 45,200 \text{ W/m}^2$

$t = 201.4 \,°C$

4. $q_1 = \dfrac{t_1 - t_2}{\frac{\delta_1}{\lambda_1} + \frac{\delta_2}{\lambda_2} + \frac{\delta_3}{\lambda_3}} = \dfrac{20 - (-10)}{\frac{0.006}{0.78} + \frac{0.008}{0.023} + \frac{0.006}{0.78}} = 82.60 \text{ W/m}^2$

$q_1 = \dfrac{t_1 - t_2}{\frac{\delta_1}{\lambda_1}} = \dfrac{20 - (-10)}{\frac{0.006}{0.78}} = 3,900 \text{ W/m}^2$

$Q = A q_1 = 0.7 \times 0.6 \times 82.6 = 34.692 \text{ W}$

$\dfrac{q_2}{q_1} = \dfrac{3,900}{82.6} = 47.215$

5. $Q = 2\pi\lambda l q = \dfrac{2\pi\lambda l(t_1 - t_2)}{\ln\frac{r_2}{r_1}} = \dfrac{2\pi \times 1 \times 0.15 \times (65 - 0)}{\ln\frac{2.5}{1.5}} = 119.8 \text{ W}$

$119.8\text{W} = I^2 R$

$I = 199.863 \text{ A}$

6. $\Phi_l = \dfrac{t_{w1} - t_{w2}}{\frac{1}{2\pi\lambda}\ln\frac{d_2}{d_1}} = \dfrac{500 - 50}{\frac{1}{2\pi \times 0.05}\ln\frac{d_2}{0.1}} = 180 \text{ W/m} \Rightarrow d_2 = 219.3 \text{ mm}$

$l = \dfrac{(d_2 - d_1)}{2} = 59.65 \text{ mm}$

7. When the small thermal conductivity of the material is close to the wall of the tube:

$$\Phi_1 = \dfrac{(t_1 - t_2)}{\frac{\ln\left(\frac{50+75}{50}\right)}{2\pi l\lambda_1} + \frac{\ln\left(\frac{50+75+75}{50+75}\right)}{2\pi l\lambda_2}} = \dfrac{2\pi l(t_1 - t_2)}{19.19}$$

When the large thermal conductivity of the material is close to the wall of the tube:

$$\Phi_2 = \dfrac{2\pi l(t_1 - t_2)}{\frac{\ln 1.6}{\lambda_1} + \frac{\ln 2.5}{\lambda_2}} = \dfrac{2\pi l(t_1 - t_2)}{15.47}$$

Therefore, when the material with large thermal conductivity is close to the pipe wall, the thermal insulation effect is better.

If it is a flat wall:

$$q = \dfrac{t_1 - t_2}{\frac{\delta_1}{\lambda_1} + \frac{\delta_2}{\lambda_2}}$$

Since $\delta = \delta_1 = \delta_2$, this problem does not exist.

Chapter 11
Convection heat transfer

11.1 Convective heat transfer

Convective heat transfer refers to the heat transfer phenomenon between the fluid and the solid surface when the fluid flows through the solid (Fig. 11.1), which is the result of the combined effect of heat convection and heat conduction.

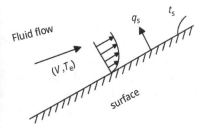

Fig. 11.1: Convective heat transfer.

Convective heat transfer is very common in life and industry. Here are a few examples: (1) heat pipes rely on liquid flow for heat exchange; (2) electronic components rely on air cooling or liquid cooling for heat dissipation; (3) heavy rainy weather often occurs in summer; this is because airflow movement occurs in the atmosphere, and thunderstorms and storms are formed due to convection in the atmosphere; (4) the movement of land and sea winds in nature, adjusting the temperature by means of heat transfer; (5) the flame burning in the furnace in the industrial furnace transmits heat to the furnace wall in two ways: convection and radiation; (6) the boiling of water in the kettle when boiling water is a convective heat transfer process in which there is a phase transition.

There are many factors that affect convective heat transfer, and in order to obtain a formula suitable for practical calculations, it is necessary to classify the heat transfer process by category. According to the presence or absence of phase transition in the heat transfer process, it can be divided into convective heat transfer with phase change and convective heat transfer without phase change. In the convective heat transfer with phase change, the main occurrence is the two phase-change process of boiling and condensation, of which boiling heat transfer can be divided into large container boiling and intratube boiling, condensation heat transfer can be divided into outside the tube condensation and intratube condensation. In convective heat transfer without phase transition, according to the different causes of flow, it can be divided into three categories: natural convection, forced convection, and mixed convection. Due to the different causes of flow, the velocity field in the fluid is also different, so the heat transfer law is also different. In natural convection, according to the size of the convective space, it

https://doi.org/10.1515/9783111329703-012

can be divided into large space natural convection and finite space natural convection. In forced convection, according to the different flow forms, it can be divided into forced convective heat transfer in the circular tube, convective heat transfer in the outer swept plate, convective heat transfer in the outer swept single circular tube, convective heat transfer in the outer swept tube cluster, jet shock heat transfer, and so on. When studying a certain type of convective heat transfer, it is also necessary to pay attention to its differences in physical processes with other convective heat transfer types, so as to better understand the differences in the calculation formula of the heat transfer process.

11.2 Newton's cooling law

For convective heat transfer processes as fluids flow through a solid surface (Fig. 11.2), convective heat transfer can be calculated using Newton's cooling formula. Newton's cooling law was determined experimentally by Newton in 1701, and it is in good compliance with reality when forced convection is good, and only when the temperature difference is not too large in natural convection. It is one of the basic laws of heat transfer.

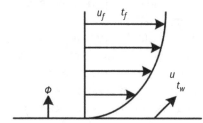

Fig. 11.2: Newton's cooling law.

Newton's cooling law:

$$\Phi = Ah(t_w - t_f) \tag{11.1}$$

$$q = h(t_w - t_f) \tag{11.2}$$

where h is the average surface heat transfer coefficient for the entire solid surface; t_w the average temperature of solid surfaces; t_f the fluid temperature; for external bypass, t_f is the mainstream temperature away from the wall; for internal flow, t_f is the average temperature of fluid.

Because the actual problems solved by convective heat transfer are often more complex, the heat transfer conditions along the solid surface (Fig. 11.3) (such as the geometric conditions of the solid surface, the surface temperature, and the flow state of the fluid) change so that the local surface heat transfer coefficient, local temperature difference, and local heat flow density will change along the solid surface. For local convective heat transfer, the Newtonian cooling formula can be expressed as

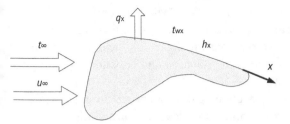

Fig. 11.3: Newton's cooling law.

$$q_x = h_x(t_w - t_f)_x \tag{11.3}$$

The total convective heat exchange heat on the entire solid surface area A can be written as

$$\Phi = \int_A q_x dA = \int_A h_x(t_x - t_f)dA = (t_w - t_f)\int_A h_x\, dA \tag{11.4}$$

If the surface temperature of the solid is uniform (equal wall temperature boundary) and the temperature difference between the wall surface and the fluid is the same everywhere, then there is

$$(t_w - t_f)_x = t_w - t_f = \text{constant} \tag{11.5}$$

Compared to eq. (11.1), the relationship between the average surface convective heat transfer coefficient and the local surface convective heat transfer coefficient under uniform solid surface temperature can be obtained:

$$h = \frac{1}{A}\int_A h_x\, dA \tag{11.6}$$

Newton's law of cooling describes the relationship between convective heat exchange and surface heat transfer coefficient and temperature difference, which is the defining formula of surface heat transfer coefficient. Although the form is simple, the difficulty lies in how to determine the surface heat transfer coefficient. Determining the size of the surface heat transfer coefficient is also a core problem of convective heat transfer.

There are four methods for studying convective heat transfer problems, that is, to obtain the expression of the surface heat transfer coefficient h:

(1) Analytical method: The analytical method is a mathematical solution to the partial differential equations and corresponding fixed solution conditions describing a certain type of convective heat transfer problem, so as to obtain the analytical solution of the velocity field and the temperature field; because the mathematical process is often more complex, there are difficulties in solving, and at present, only a few relatively simple analytical solutions to the convective heat transfer problem can be obtained. However, due to the fact that in this method, the mathematical analysis

process is more rigorous, the physical concepts and logical reasoning are relatively clear, and the solution results can be expressed in the form of functions, which can profoundly reveal the dependence relationship of each physical quantity on the surface heat transfer coefficient, and is the standard and basis for evaluating the results obtained by other methods.

(2) Experimental method: At present, in the engineering design, the main basis for obtaining the surface heat transfer coefficient calculation formula is still the experiment in order to reduce the number of experiments and improve the versatility of the experimental measurement results, the experimental determination of heat transfer should be carried out under the guidance of similar principles. With the advancement of modern measurement technology, observations of the fine structure and phenomena of convective heat transfer that were previously impossible can now be achieved. And the experimental method plays a key role in the exploration of the unsolved convective heat transfer mechanism (boiling heat transfer). Due to the increased accuracy of measurements, experimental results are also commonly used to test the accuracy of other methods.

(3) Analogy method: The analogy method is to use the commonality of heat transfer and momentum transfer in the mechanism so as to establish the analogy relationship between the surface heat transfer coefficient and the friction coefficient and derive the heat transfer coefficient of the convective heat exchange surface through the friction coefficient data obtained by the fluid flow experiment that is relatively easy to carry out. In the early days of heat transfer, this method was widely used to obtain the formula for the calculation of turbulent heat transfer processes. With the rapid development of experimental testing technology and computer technology, the application of this method has become less and less in recent years, but the similarity of the mechanism between momentum transfer and heat transfer based on this method is very helpful for understanding and analyzing the convective heat transfer process.

(4) Numerical method: In the past four decades, the numerical solution method of convective heat transfer has been rapidly developed and has increasingly shown its important role. In the previous chapter of the study of thermal conduction problems, numerical methods were used to solve them. Since the convective heat transfer equation is more complex, the numerical solution of convective heat transfer has added two difficulties, that is, the discrete convection term and the numerical processing of the pressure gradient term in the momentum equation, the solution of these two difficulties is more complicated, involving many specialized numerical methods.

At present, when solving complex convective heat transfer problems, scientific and technological workers often use a combination of theoretical analysis, numerical calculation, and experimental research.

11.3 Influence factors on convective heat transfer

Convective heat transfer is the result of the combined effect of two basic heat transfer modes: fluid heat conduction and heat convection. Therefore, all factors that affect fluid heat conduction and heat convection will affect convective heat transfer. There are five main sections:

11.3.1 Cause of flow

Due to the different causes of the flow, the velocity field and temperature field distribution in the flow body are different, resulting in different convective heat exchange laws.

There are natural convective heat transfer and forced convective heat transfer. Convection is called forced convection if the fluid is forced to flow over the surface by external means such as a fan, pump, or the wind. In contrast, convection is called natural convection if the fluid motion is caused by buoyancy forces that are induced by density differences due to the variation of temperature in the fluid. One of the most typical examples of natural convection is the flow of air around the indoor radiator, where the heating temperature of the air around the radiator increases, the density changes, and the hot air flows upward.

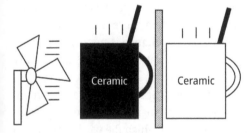

Fig. 11.4: Forced and natural convective heat transfer.

Figure 11.4 shows an example of natural convection versus forced convection in life. The figure on the left uses a fan to accelerate the flow of air, thereby accelerating the heat loss of hot water, which is forced convection. The natural cooling of hot water placed in the space in the picture on the right belongs to natural convection.

Generally speaking, the velocity of natural convection is low, so natural convective heat transfer is usually weaker than forced convective heat transfer, and the surface heat transfer coefficient is smaller. For example: natural convective h of gas is 1–10 $W/(m^2 \cdot K)$, and forced convective h of gas is 10–100 $W/(m^2 \cdot K)$. Forced convection can speed up heat transfer and is widely used in engineering.

11.3.2 Flow state

The state of the flow can be laminar flow and turbulence flow or the flow between them, which is called transitional flow. When the flow rate is very small, the fluid flows in layers and does not mix with each other, called laminar flow, or sheet flow; gradually increase the flow rate, the streamline of the fluid begins to appear wavy swing, the frequency, and amplitude of the swing increase with the increase of the flow rate, this flow condition is called transitional flow; when the flow rate increases to a large size, the streamline is no longer clearly discernible, and there are many small swirls in the flow field, called turbulence.

Laminar: The flow velocity is slow, and the fluid flows in layers parallel to the wall surface. The heat transfer in the direction perpendicular to the flow mainly depends on molecular diffusion (i.e., heat conduction).

Turbulence: There are strong pulsations and vortices in the fluid, which make the various fluids mix rapidly. Heat transfer relies on molecular diffusion and turbulent pulsation, so turbulent convective heat transfer is stronger than laminar convective heat transfer, and the surface heat transfer coefficient is larger.

11.3.3 Whether the fluid has a phase change

When there is no phase change in the fluid, convective heat transfer is caused by the exchange and change of the sensible heat of the fluid, when there is a phase change in the fluid, the heat exchanged by the convective heat exchange is mainly caused by latent heat, and the heat transfer law changes. Phase transitions in fluids are mainly divided into two categories: boiling and condensation.

(1) Boiling heat transfer: Boiling is a violent vaporization process in which a large number of bubbles form inside the working medium and change from liquid to gaseous. Boiling heat transfer refers to a heat transfer method in which the working medium takes away the heat through the movement of air bubbles and cools it. Phase change latent heat is absorbed when boiling.

(2) Condensation heat exchange: Condensation heat transfer is the heat transfer of steam to the solid wall during the condensation process. It can be divided into membranous coagulation and bead-like coagulation. Phase change latent heat is released during condensation.

11.3.4 Physical properties of fluid

The physical properties of the fluid have a great influence on convective heat transfer because convective heat transfer is the result of the joint action of two basic heat transfer methods of heat conduction and convection, so the physical properties that affect heat conduction and convection will affect convective heat transfer. Taking forced convective heat transfer without phase transition as an example, the density, dynamic viscosity, volume coefficient, and specific constant pressure heat capacity of the fluid will affect the distribution of velocity and heat transfer in the fluid, thus affecting the convective heat transfer. If the cooling medium of the internal cooling generator is changed from air to water, the output of the generator can be increased, which is to take advantage of the fact that the thermal physical properties of water are conducive to strengthening convective heat transfer. The main physical properties involved in the convective heat transfer analysis and their respective effects are as follows:

1) Thermal conductivity λ, W/(m^2 · K): λ is larger, the smaller the thermal resistance of the fluid, and the stronger the convective heat transfer.

    (1) Density ρ, kg/m^3 and heat capacity c, J/(kg · K): ρv reflects the heat capacity of a unit volume of fluid. The larger the value, the more heat transferred through convection, and the stronger the convective heat transfer.

    (2) Dynamic viscosity η, Pa · s; kinematic viscosity $v = \eta/\rho$, m^2/s: This affects velocity distribution and flow pattern and therefore convective heat transfer.

    (3) Body expansion coefficient a_v, K^{-1}: This affects the buoyancy of the fluid in the gravity field due to the density difference and therefore affects natural convective heat transfer.

The physical properties of the fluid vary not only with the type, temperature, and pressure of the fluid but also have a great relationship with the temperature for the same incompressible Newtonian fluid. When analyzing the physical phenomena of convective heat transfer, numerical values at qualitative temperatures are usually used. Qualitative temperature refers to the temperature used by a medium to find mutual property parameters. For different flow fields, the choice of qualitative temperature is different. The external flow often chooses the arithmetic average of the incoming fluid temperature and the solid wall temperature, and the internal flow often chooses the average (arithmetic or logarithmic average) of the fluid inlet and outlet temperature in the tube, with exceptions.

11.3.5 Geometrical factors of heat transfer surface

Geometrical factors such as the geometry, size, relative position, and surface roughness of the heat transfer surface will affect the flow state of the fluid and therefore affect the velocity and temperature distribution of the fluid and affect convective heat transfer. For example, the forced convection flow in the tube is completely different from the forced convection of the fluid across the circular tube, the former is the flow in the tube, which belongs to the scope of internal flow, and the latter is the flow of the external swept object, which belongs to the scope of the external flow. The conditions of these two flows are different, so their heat transfer laws are different. For another example, for horizontal walls, the flow of heat dissipation is completely different in the two cases of heat side up arrangement and heat side down arrangement, so the heat exchange law is also different.

In summary, there are many factors affecting convective heat transfer. Table 11.1 shows h-range of some convective heat transfer processes, and the surface heat transfer coefficient h is a function of many variables:

$$h = f\left(u, t_w, t_f, \lambda, \rho, c, \eta, a, l, \psi\right) \tag{11.7}$$

where l is the feature length and ψ the geometric factor.

One of the purposes of studying convective heat transfer is to determine the specific expressions of the surface heat transfer coefficients under different heat transfer conditions. The main methods include analytical methods, numerical methods, experimental methods, and analog methods. The mathematical description of convective heat transfer and the derivation of the formula for calculating the heat transfer coefficient on its surface are self-taught.

Tab. 11.1: The h-range of some convective heat transfer processes.

Type of convection	Convection coefficients h $(W/(m^2 \cdot K))$
Air natural convection	1–10
Air forced convection	10–100
Water natural convection	100–1,000
Water forced convection	1,000–15,000
Water boiling	2,500–35,000
Water vapor condensation	5,000–25,000

11.4 Summary

Focus on the following:
(1) Newton's cooling law
(2) Influence factors on convective heat transfer
(3) Numerical concept of convective heat transfer coefficient

Exercises

1. Convection is the mode of energy transfer between a solid surface and adjacent liquid or gas, which is ( ).
   A. True   B. False
2. Convective heat transfer involves the combined effects of conduction and radiation, which is ( ).
   A. True   B. False

3. The convention heat transfer coefficient h is a property of the fluid, which is ( ).
   A. True   B. False
4. The cooling of the hot bread in Fig. 11.5 includes ( ).

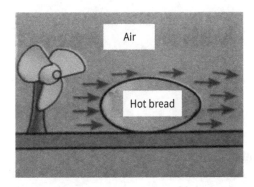

Fig. 11.5: Exercise 4: attached graph.

   A. conduction   B. natural convection   C. forced convection   D. radiation
5. What is the unit of the convention heat transfer coefficient h?
   A. W/m^2 B. $W/(m^2 \cdot K)$ C. W/K D. $W \cdot K/m^2$
6. Which factor will not affect convective heat transfer?
   A. Cause of flow   B. Flow state   C. Phase change   D. Emissivity of the surface
7. What's the definition of forced convention and natural convention?
8. In the movie Titanic, the hero Jack is frozen to death in the sea, while the heroine Rose survives by lying on the raft. Try to explain this phenomenon from the perspective of heat transfer.
9. In winter, under the same outdoor conditions, why does it feel colder when there is wind than when there is no wind?
10. Is the natural convective heat transfer experiment involved in a laboratory on the Earth's surface still valid in space? Why?
11. Hot air at 90 °C is blown over a 2.5 m × 4 m flat surface at 30 °C. If the average convention heat transfer coefficient is 45 $W/m^2 \cdot °C$, determine the rate of heat transfer from the air to the plate, in kW.
12. A 2.5-m long, 0.2-cm-diameter electrical wire extends across a room at 22 °C. Heat is generated in the wire as a result of resistance heating, and the surface temperature of the wire is measured to be 142 °C in steady operation. Also, the voltage drops and electric current through the wire is measured to be 80 V and 1.2 A, respectively. Disregarding any heat transfer by radiation, determine the heat loss from the wire.
13. Base on question 7, determine the convention heat transfer coefficient for heat transfer between the outer surface of wire and the air in the room.
14. A horizontally placed steam pipe with an insulation outer diameter of $d = 583$ mm and an average measured temperature of 48 °C on the outer surface. The air temperature is 23 °C, and the surface heat transfer coefficient $h = 3.42$ $W/(m^2 \cdot K)$ for

natural convective heat transfer between air and the outer surface of the pipe. Calculate the natural convection heat exchange on pipes of each meter length.

Answers

1. A
2. B
3. B
4. A, C, D
5. B
6. D
7. Convention is called forced convention if the fluid is forced to flow over the surface by external means such as a fan, pump, or the wind. In contrast, convention is called natural convention if the fluid motion is caused by buoyancy forces that are induced by density differences due to the variation of temperature in the fluid.
8. Jack exchanges heat between his body and seawater in seawater due to natural convection, while Rose creates natural convection between his body and air on the raft. All other things being equal, the natural convection of water is much stronger than that of air, so Jack's body loses energy much faster than Rose due to natural convection. So Jack froze to death while Rose survived.
9. Assuming that the surface temperature of the human body is the same, the heat dissipation of the human body is equivalent to forced convective heat exchange when there is wind, and it belongs to natural convective heat exchange when there is no wind. The forced convective heat transfer of the air is stronger than the natural convection, so the human body takes away more heat when there is wind, so it feels colder.
10. The experiment cannot obtain the results of the experiment on the ground in space. Because natural convection is due to the temperature difference inside the fluid, which causes a difference in density and is caused by gravity, the laboratory device in space will be in a state of weightlessness, so natural convection cannot be formed, so the expected experimental results cannot be obtained.
11. $Q = hA(T_1 - T_2) = 45 \dfrac{W}{m^2 \cdot K} \times (2.5\,m \times 4\,m) \times (90 - 30)\,K = 27\,kW$
12. When steady operating conditions are reached, the heat loss from the wire equals the rate of heat generation in the wire as a result of resistant heating. That is,

$$Q = E_{generate} = VI = (80\,V) \times (1.2\,A) = 96\,W$$

13. The surface area if the wire is

$$A_s = \pi DL = \pi(0.002\,m) \times (2.5\,m) = 0.01571\,m^2$$

Newton's cooling law of convention heat transfer is expressed as

$$Q_{conv} = hA_s(T_s - T_\infty)$$

Disregarding any heat transfer by radiation and thus assuming all the heat loss from the wire to occur by convention, the convention heat transfer coefficient is determined to be

$$h = \frac{Q_{conv}}{A_s(T_s - T_\infty)} = \frac{\frac{96\,W}{0.01571\,m^2}}{142 - 22}\,°C = 50.9\,W/(m^2 \cdot K)$$

14. Natural convective heat dissipation per unit length is

$$Q = \pi dh(t_w - t_f) = 3.14 \times 0.583\,m \times 3.42\,W/(m^2 \cdot K) \times (48\,°C - 23\,°C) = 156.5\,W/m$$

Chapter 12
Radiative heat transfer

12.1 Basic concepts of thermal radiation

Heat is transferred by means of electromagnetic waves (radiant energy or light) when a hot body emits radiation because of temperature (e.g., as a result of vibrational and rotational motions of molecules, atoms, and electrons of a substance.) Radiative heat transfer refers to the total effect of mutual radiation and absorption between objects. When an object is in thermal equilibrium with the environment, the thermal radiation on its surface is still continuous, but its net-radiated heat transfer is equal to zero. Compared with thermal conduction and convection, thermal radiation has two differences: (1) the energy transfer of thermal radiation does not need to rely on a medium, and the transfer efficiency is the highest in vacuum; (2) the conversion of energy forms occurs during the process of emitting and absorbing energy by the object (electromagnetic energy and thermal energy).

When radiation strikes a surface, part of it is absorbed, part of it is reflected, and the remaining part, if any, is transmitted (Fig. 12.1).

12.1.1 Absorption

Definition of input radiation: Radiated energy in a full wavelength range projected onto the surface of a unit area per unit time, G, W/m^2. The rest can be done in the same manner; for example, the representative letter for the absorption of radiation is G_a and the unit is W/m^2. The representative letter for the reflective radiation is G_ρ and the unit is W/m^2. The representative letter for the transmitted radiation is G_τ and the unit is W/m^2. a, ρ, τ are called the absorption ratio, reflection ratio, and transmission ratio of the projected radiant energy by the object. They are represented as follows:

The absorption ratio:

$$\alpha = \frac{G_a}{G} \tag{12.1}$$

The reflection ratio:

$$\rho = \frac{G_\rho}{G} \tag{12.2}$$

The transmission ratio:

$$\tau = \frac{G_\tau}{G} \tag{12.3}$$

https://doi.org/10.1515/9783111329703-013

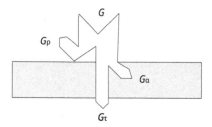

Fig. 12.1: Energy conservation.

According to energy conservation,

$$G_a + G_p + G_\tau = G \tag{12.4}$$

The following conclusions can be drawn:

$$\alpha + \rho + \tau = 1 \tag{12.5}$$

If the input radiation is at a certain wavelength of λ and radiation energy G_λ, the parts absorbed, reflected, and transmitted by the object are $G_{\lambda a}$, $G_{\lambda \rho}$, $G_{\lambda \tau}$, then we have:

The spectral absorption ratio:

$$\alpha_\lambda = \frac{G_{\lambda a}}{G_\lambda} \tag{12.6}$$

The spectral reflection ratio:

$$\rho_\lambda = \frac{G_{\lambda \rho}}{G_\lambda} \tag{12.7}$$

The spectral transmission ratio:

$$\tau_\lambda = \frac{G_{\lambda \tau}}{G_\lambda} \tag{12.8}$$

$$\alpha_\lambda + \rho_\lambda + \tau_\lambda = 1 \tag{12.9}$$

So, we can figure out the relationship between α, ρ, τ and α_λ, ρ_λ, τ_λ:

$$\alpha = \frac{\int_0^\infty \alpha_\lambda G_\lambda d\lambda}{\int_0^\infty G_\lambda d\lambda} \tag{12.10}$$

$$\rho = \frac{\int_0^\infty \rho_\lambda G_\lambda d\lambda}{\int_0^\infty G_\lambda d\lambda} \tag{12.11}$$

$$\tau = \frac{\int_0^\infty \tau_\lambda G_\lambda d\lambda}{\int_0^\infty G_\lambda d\lambda} \tag{12.12}$$

Moreover, here are still some we need to be aware of:

First, α_λ, ρ_λ, τ_λ are spectral radiation characteristics of an object that depends on the type, temperature, and surface condition of the object and is a function of wavelength. But α, ρ, τ depend not only on the nature of the object but also on the wavelength distribution of the radiation energy projected.

Second, in fact, when thermal radiation is projected onto the surface of a solid or liquid, part of it is reflected and the rest is completely absorbed within the surface layer. The thickness of this surface layer is very thin (metal: the thickness of the surface layer is less than 1 μm; most non-metals: the thickness of the surface layer is less than 1 mm).

Therefore, it can be considered that the transmittance ratio of solids and liquids to thermal radiation is zero:

$$\tau = 0, \alpha + \rho = 1 \qquad (12.13)$$

12.1.2 Reflection

Reflection is divided into specular reflection and diffuse reflection. Specular reflection is characterized by "reflection angle = incident angle." However, diffuse reflection is characterized by "reflected radiation can be distributed evenly in all directions" (Fig. 12.2).

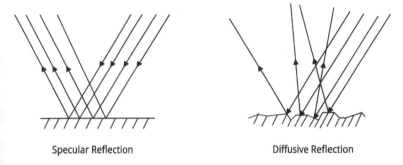

Specular Reflection   Diffusive Reflection

Fig. 12.2: Contrast of specular versus diffuse reflection.

The kind of reflection depends on the roughness of the surface of the object and the wavelength of the projected radiant energy. When the scale of roughness is smaller than the wavelength of the projected radiant energy, specular reflection occurs; otherwise diffuse reflection occurs. The reflection of thermal radiation in most engineering materials is similar to diffuse reflection.

12.1.3 Gray body and blackbody

Different objects in nature are very different, which has caused great difficulties in the study of thermal radiation problems. For the convenience of research, the concept of gray body and blackbody was introduced.

The gray body is an imaginary object whose spectral radiation characteristics do not change with wavelength. a_λ, ρ_λ, τ_a are all constants:

$$a_\lambda = a, \quad \rho_\lambda = \rho, \quad \tau_\lambda = \tau \tag{12.14}$$

The absorption ratio, reflectance, and transmission ratio of the gray body are equal to the spectral absorption ratio, spectral reflectance, and spectral transmission ratio, respectively, and the numerical size has nothing to do with the wavelength, but only depends on the nature of the gray body itself. In the study of practical problems, the vast majority of objects can be treated as gray bodies.

Absolute blackbody is an object with an absorption ratio $\alpha = 1$ and is referred to as blackbody for short. Blackbody, like gray body, is an ideal object. Blackbody is able to absorb all the electromagnetic radiation from outside and will not have any reflection and transmission. In other words, the blackbody has an absorption coefficient of 1 and a transmission coefficient of 0 for electromagnetic waves of any wavelength. A blackbody does not have to be black (e.g., the sun can be regarded as a blackbody in some cases), even if it cannot reflect any electromagnetic waves, it can also emit electromagnetic waves, and the wavelength and energy of these electromagnetic waves depend on the temperature of the blackbody, not changed by other factors. True blackbody does not exist in nature, but many objects are better blackbody approximations (on some electromagnetic bands).

Besides, mirror body (called white body in diffuse reflection): $\rho = 1$. Absolutely transparent body is $\tau = 1$. Obviously, mirror body and absolute transparent body are idealized objects that do not exist in nature like blackbody.

12.1.4 Radiation intensity

In order to illustrate the distribution of radiation energy emitted by the surface of an object in all directions of space, the concept of radiation intensity is introduced. Definition of radiation intensity: the radiant energy of all wavelengths contained in the cubic angle of the unit from the unit projected area (visible area) in a unit time (Fig. 12.4). Radiation intensity indicates how much radiant energy the surface of an object emits in a certain direction in space.

The concept of solid angles is introduced first (Fig. 12.3). Definition of solid angle: The area A of the spherical surface with a radius r and the center of the sphere, the unit is Sr (Steradian):

$$\Omega = \frac{A}{r^2} \tag{12.15}$$

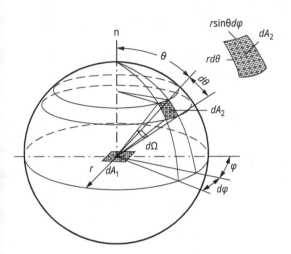

Fig. 12.3: Schematic diagram of solid angle.

On the sphere with a microstructure area of dA_1 and a radius of r above, there is a microstructure area dA_2 cut by the longitude and latitude lines in the direction (θ, φ). The micro-element stereoangle of dA_2 for the center of the sphere can be expressed as

$$d\Omega = \frac{dA_2}{r^2} = \frac{rd\theta \cdot r\sin\theta d\varphi}{r^2} = \sin\theta d\theta d\varphi \tag{12.16}$$

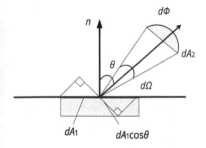

Fig. 12.4: Schematic diagram for radiation intensity.

If the radiation energy emitted by the microstructure area dA_1 to dA_2 per unit time is $d\varphi$, and the projection area of dA_1 in the direction of θ is $dA_1\cos\theta$, the radiation energy emitted by the unit projection area can be expressed as

$$L(\theta, \varphi) = \frac{d\varphi}{dA_1\cos\theta d\Omega} \tag{12.17}$$

$L(\theta, \varphi)$ is called radiation intensity of dA_1 in the direction of (θ, φ) or directional radiation intensity with the unit of W/(m^2 · Sr).

Radiation intensity depends not only on the type of object, surface properties, and temperature, but also on direction. For the surface of an isotropic object, the radiation intensity is free from angle φ, $L(\theta, \varphi) = L(\theta)$.

The radiation intensity for a wavelength of radiation energy is called spectral radiation intensity. Relationship between radiation intensity and spectral radiation intensity is

$$L(\theta) = \int_0^\infty L_\lambda(\theta)d\lambda \tag{12.18}$$

Depending on whether the unit used is meter or micrometer depending on the wavelength, the unit of spectral radiation intensity is W/(m^3 · Sr) or W/(m^2 · µm · Sr).

12.1.5 Emissive power

The radiation energy of the entire wavelength emitted from the surface of each unit area to hemispheric space in a unit time. This is represented in E, and its unit is W/m^2.

12.1.5.1 Spectral emissive power

A specific wavelength of radiant energy emitted into hemispheric space by the surface of a surface per unit area into hemispheric space in a unit time, represent in E_λ, unit is W/m^3. The relationship between E and E_λ can be expressed as

$$E = \int_0^\infty E_\lambda d\lambda \tag{12.19}$$

12.1.5.2 Directional emissive power

In unit time, the radiation energy in the three-dimensional angle of the unit area surface emitted in a certain direction is represented in E_θ, unit is W/(m^2 · Sr).

Compare their definitions, the relationship between emissive power and directional emissive power:

$$E = \int_{\Omega = 2\pi} E_\theta d\Omega \tag{12.20}$$

Relationship between directional emissive power and radiation intensity:

$$E_\theta = L(\theta)\cos\theta \tag{12.21}$$

Relationship between emissive power and radiation intensity:

$$E = \int_{\Omega=2\pi} L(\theta)\cos\theta d\Omega \tag{12.22}$$

12.2 The basic law of blackbody radiation

12.2.1 Planck's law

Max Planck (1858–1947), German theoretical physicist, founder of quantum mechanics, and one of the most important physicists of the twentieth century, made progress in physics by discovering the energy quantum. He made important contributions and won the Nobel Prize in physics in 1918. In 1900, Max Planck gave the relationship between the blackbody emissive power $E_{b\lambda}$, the absolute temperature T, and wavelength λ based on quantum assumption, called Planck's law:

$$E_{b\lambda} = \frac{C_1 \lambda^{-5}}{e^{C_2/(\lambda T)} - 1} \tag{12.23}$$

C_1 is Planck's first constant and C_2 is Planck's second constant; $C_1 = 3.743 \times 10^{-16}$ W·m^2 and $C_2 = 1.439 \times 10^{-2}$ m·K.

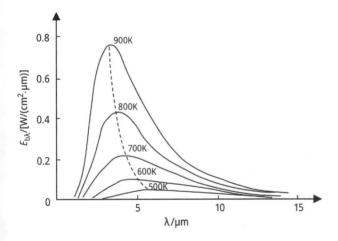

Fig. 12.5: Blackbody radiation curve.

The spectral radiation force of blackbody changes with wavelength at different temperatures as shown in Fig. 12.5. As shown in the figure, Planck's law has some characteristics, for example, the higher the temperature, the greater the spectral radiant power at the same wavelength; at a certain temperature, the spectral radiation of the black body has the maximum value at a certain wavelength; as the temperature rises, the position of λ_{max} (where $E_{b\lambda}$ is largest) downshifts, that is, the position of λ in the coordinates moves in the direction of the short wave.

12.2.2 Wien's displacement law

As temperature increases, the blackbody curves get steeper and shifts to the left to the shorter wavelength region. This law calls Wien's displacement law. The wavelength λ_{max} corresponding to the maximum spectral radiation force has the following relationship with temperature T:

$$\lambda_{max}T = 2.8976 \times 10^{-3} \approx 2.9 \times 10^{-3} \, \text{m} \cdot \text{K} \tag{12.24}$$

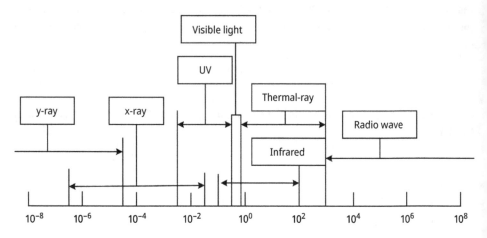

Fig. 12.6: Solar radiation energy.

Historically, the discovery of Wien's displacement law preceded Planck's law, but eq. (12.24) can be derived by deriving from λ in eq. (12.23) and making it equal to 0. According to the Wayne's law of displacement, it is possible to determine the wavelength corresponding to the maximum spectral radiation force of a blackbody at any temperature. For example, the surface temperature of sun is 5,800 K can be obtained from the above formula: λ_{max} = 0.5 μm, located within the visible light range (0.38–0.76 μm), as shown in Fig. 12.6, visible light accounts for about 44.6% of solar radiation energy.

12.2.3 Stefan–Boltzmann's law

Stefan–Boltzmann's law establishes the relationship between the blackbody emissive power $E_{b\lambda}$ and the absolute temperature T. It was first obtained from the experiment in 1879 and then by Boltzmann in 1884 using the theory of thermodynamics.

$$E_b = \sigma T^4 \tag{12.25}$$

where $\sigma = 5.67 \times 10^{-8}\ \text{W/(m}^2 \cdot \text{K}^4)$ is called the Stefan–Boltzmann constant, also known as the blackbody radiation constant.

This law, also known as the law of the four squares of radiation, is the basis for the calculation of thermal radiation engineering. It indicates that as the temperature rises, the radiation force increases dramatically.

12.2.4 Lambert's law

Lambert's law gives the law of the distribution of blackbody-radiated energy in spatial directions. Theoretically, it can be proved that the radiation intensity of the blackbody is independent of the direction, and the radiation intensity in all directions of the hemisphere space is equal. The law of uniform spatial distribution followed by such blackbody radiation intensity is called Lambert's law.

Definition of diffuse emitter: An object with equal radiation intensity in all directions of space. The blackbody is also a diffuse emitter. For diffuse emitter:

$$L(\theta) = L = \text{constant} \tag{12.26}$$

According to the relationship between directional emissive and radiation intensity:

$$E_n = L(\theta)\cos\theta = L\cos\theta = E_n\cos\theta \tag{12.27}$$

E_n is directional emissive in the direction of the surface normal.

This equation shows that the energy radiated out per unit area of the blackbody is unevenly distributed in different directions of space, changing according to the cosine law of the latitude angle of space θ: the direction perpendicular to the surface is greatest, and the direction parallel to the surface is 0. Hence, Lambert's law is also known as cosine's law.

For diffuse emitters, according to the relationship between emissive power E and radiation intensity L, it can be obtained:

$$E = \int_{\Omega=2\pi} L(\theta)\cos\theta d\Omega$$

$$= \int_{\Omega=2\pi} L\cos\theta \sin\theta d\theta d\varphi$$

$$= L \int_0^{2\pi} d\varphi \int_0^{\pi/2} \sin\theta\cos\theta d\theta = \pi L$$

That is, the radiation force of the diffuse emitter is π times the radiation intensity.

12.3 Emission characteristics of actual objects

12.3.1 Thermal radiation of actual objects

As pointed out earlier, the blackbody is the standard object for the study of thermal radiation, and the radiation characteristics of the actual object will be studied on the basis of comparison with the blackbody radiation characteristics. Since the actual object cannot fully absorb the radiation energy invested on its surface, the radiation force of the actual object is always less than the radiation of the blackbody at the same temperature. In order to illustrate the emission characteristics of actual objects, the concept of emissivity is introduced. Definition of emissivity (blackness): the ratio of the emissive power of the actual object to the emissive power of the blackbody at the same temperature:

$$\varepsilon = \frac{E}{E_b} \tag{12.29}$$

Emissivity reflects the ability of an object to emit radiant energy.

The ratio of the spectral radiation force of the actual object to the spectral radiation force of the blackbody at the same temperature is called the spectral emissivity of the object (or the spectral blackness). The expression of spectral emissivity (spectral blackness):

$$\varepsilon_\lambda = \frac{E_\lambda}{E_{b\lambda}} \tag{12.30}$$

The relationship between emissivity and spectral emissivity:

$$\varepsilon = \frac{\int_0^\infty \varepsilon_\lambda E_{b\lambda} d\lambda}{E_b} \tag{12.31}$$

But for gray body, spectral radiation characteristics do not vary with wavelength, ε_λ = constant:

$$\varepsilon = \frac{\varepsilon_\lambda \int_0^\infty E_{b\lambda} d\lambda}{E_b} = \varepsilon_\lambda \tag{12.32}$$

The size of the emissivity of an object is only related to the object itself that emits radiation and does not involve external conditions. The emissivity of different types of substances varies and generally depends on the type of substance, surface temperature, and surface condition. Here are a few specific examples to illustrate: (1) the emissivity of the steel sheet with a smooth oxide skin at room temperature is 0.82, while the emissivity of the galvanized iron sheet is only 0.23; (2) the emissivity of the polished aluminum surface is 0.04 and 0.06 at the temperature of 50 and 500 °C, respectively; (3) for metal materials, the emissivity of highly polished surfaces is very small, while the emissivity of rough surfaces or oxidized surfaces is often several times that of polished surfaces. The emissivity of most non-metallic materials is high and has little relation to the surface condition.

12.3.2 G. R. Kirchhoff's law

Kirchhoff's law reveals the relationship between an object's ability to absorb radiant energy and its ability to emit radiant energy. Its expression is

$$a_\lambda(\theta, \varphi, T) = \varepsilon_\lambda(\theta, \varphi, T) \tag{12.33}$$

It shows that the stronger the ability to absorb radiant energy, the stronger the ability to emit radiant energy. Among objects with the same temperature, the blackbody has the strongest ability to absorb radiant energy and the strongest ability to emit radiant energy.

For diffusers, the radiation characteristics are independent of the direction:

$$a_\lambda(T) = \varepsilon_\lambda(T) \tag{12.34}$$

For diffusive gray body, the radiation characteristics are independent of the direction and wavelength:

$$a(T) = \varepsilon(T), \qquad \varepsilon = \varepsilon_\lambda, \qquad a = a_\lambda \tag{12.35}$$

For the common temperature range ($T \le 2,000$ K) in engineering, most of the radiant energy is in the infrared wavelength range. Most engineering materials can be approximated as diffuse emission and gray body. Determine the value of the absorption ratio without causing large errors. Therefore, the emissive power of the actual object is

$$E = \varepsilon E_b = \varepsilon\sigma T^4 \tag{12.36}$$

The amount of radiant heat transfer between two objects is

$$q = E_2 - E_1 = \varepsilon\sigma\left(T_2^4 - T_1^4\right) \tag{12.37}$$

However, when studying the absorption of solar energy by the surface of an object and its own thermal radiation in the use of solar energy, one cannot mistakenly believe that the absorption ratio of solar energy = the emissivity of self-radiation.

Because nearly 50% of solar radiation is in the wavelength range of visible light, and the object's own thermal radiation is in the infrared wavelength range, the spectral absorption ratio of the actual object is selective to the wavelength of the input radiation. The emissivity of self-radiation is quite different.

For example, the selective surface coating material on solar collectors can absorb solar energy up to 0.9, while the emissivity is only 0.1. This characteristic of the material can not only absorb more solar energy but also reduce its own radiation heat dissipation loss.

12.4 Atmospheric greenhouse effect and greenhouse effect

The glass greenhouses and vegetable greenhouses that we can see in our daily life are typical greenhouses. Ordinary glass has a large penetration ratio to visible light, and infrared radiation $\lambda < 3$ μm, but the penetration of infrared radiation ($\lambda > 3$ μm) is relatively small. The visible light radiation emitted by the sun can enter the room through the glass window, while the long-wave infrared radiation emitted by indoor objects at normal temperature cannot be transmitted through the glass window, so the effect of a greenhouse is formed as shown in Fig. 12.7.

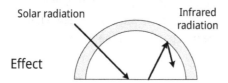

Solar radiation   Infrared radiation

Effect

Fig. 12.7: Greenhouse effect.

Any object in the universe radiates electromagnetic waves outward. The warmer the object, the shorter the wavelength of the radiation. The Sun's surface temperature is about 5,900 K, and it emits short-wavelength electromagnetic radiation called solar shortwave radiation (which includes visible light ranging from violet to red). While the ground is warmed by the Sun's short-wave radiation, it also radiates electromagnetic waves outward to cool it all the time. The electromagnetic wavelengths emitted by the earth are longer because of the lower temperature and are called terrestrial long-wave radiation. Short-wave radiation and long-wave radiation encounter different things as they pass through Earth's atmosphere: the atmosphere is nearly transparent to the Sun's short-wave radiation, but strongly absorbs ground-based long-wave radiation. While the atmosphere absorbs long-wave radiation from the ground, it also emits long-wave radiation with longer wavelengths (because the atmosphere is cooler than the ground). The part that goes down to the ground is called inverse radiation. After the ground receives reverse radiation, it will heat up, or the atmosphere has a thermal insulation effect on the ground. As a result of thermal equilibrium, the surface temperature of the earth changes between about 250 and 320 K throughout the year.

Not every gas in the atmosphere strongly absorbs terrestrial long-wave radiation. The gases that play a role in greenhouse gases in the earth's atmosphere are called greenhouse gases, mainly carbon dioxide, methane, ozone, nitrous oxide, freon, and water vapor. They absorb almost all the long-wave radiation emitted by the ground, with only a narrow section of which absorbs very little, hence the name "window area." It is mainly through this window area that the earth returns 70% of the heat obtained from the sun to the cosmic space in the form of long-wave radiation, thereby maintaining the ground temperature unchanged. In recent years, with the development of industrialization in various countries around the world, a large amount of industrial exhaust gas and automobile exhausts have been discharged into the air, which has increased the content of gases such as CO_2, SO_2, and nitrogen oxides, due to their strong absorption of infrared radiation on the surface of the earth. Function to reduce the heat radiated from the earth to space, forming the so-called atmospheric greenhouse effect, and increasing the temperature of the earth's surface.

The Intergovernmental Panel on Climate Change notes that human activities have caused about 1.0 °C of global warming since the Industrial Revolution, and that if it continues to heat at its current rate, global warming could reach 1.5 °C between 2030 and 2052. Global average temperatures are rising at an unprecedented rate, the likelihood of global warming being controlled below 1.5 °C is rapidly decreasing, and the risk of humanity crossing the irreversible tipping point of the climate system is increasing.

The intensification of the greenhouse effect will bring a variety of serious consequences, which will seriously affect the normal production and life of human beings: (1) abnormal climate. Small changes in Earth's surface temperature can trigger other changes, such as changes in atmospheric cloudiness and circulation. Some of these changes will exacerbate ground warming, while others will slow it down. Many undetermined factors also affect scientists' inferences about further trends in Earth's temperature. What is now certain is that the growing greenhouse effect is one of the causes of many extreme weather hazards; (2) increase in pests and diseases. As the global temperature rises, the Arctic ice begins to melt, and the virus that has been frozen for more than 100,000 years may reappear, causing a fatal blow to mankind; (3) sea level rise. Global warming is causing the ice sheets to melt in the Arctic and Antarctic, and the effects of rising sea levels on island nations and low-lying coastal areas are obvious. It is estimated that if the sea level rises by 1 m, 120,000 m^2 of land in China will be submerged and 70 million people will be displaced; (4) land desertification: According to the United Nations Environment Program, in the southern part of the Sahara Desert, the desert expands by about 1.5 million hectares every year. Every year, 6 million hectares of land in the world are desertified. This costs agricultural production 26 billion dollars annually.

The future increase of the earth's temperature depends on the control of human activities on the impact of climate change, that is, whether the emission of greenhouse gases can be effectively controlled. Therefore, countries around the world have issued calls for political action to enter a climate emergency and at the same time accelerate industrial, economic, and social green. Carbon transition, controlling greenhouse gas emissions and achieving carbon peaking, and carbon neutralization have become an inevitable choice for slowing global warming.

Exercises

1. A blackbody with a diameter of D and an inner wall blackness of ε, temperature of T. There is a hole with a diameter of d ($d \ll D$) on the wall of the ball. The energy radiated from the cavity through the hole is____.
2. A surface is approximately gray, then the surface is____.
   A. $\varepsilon_\lambda \neq \varepsilon$
   B. $\varepsilon_\lambda \neq \varepsilon = $ constant
   C. $\varepsilon_\lambda = \varepsilon \neq $ constant
   D. $\varepsilon_\lambda = \varepsilon = $ constant
3. The emission force of the blackbody at 1,000 °C is_____.
4. There is a blackbody with a temperature of 3,527 °C and the longest wavelength of monochromatic emission force is_____.
5. The temperature of the blackbody surface increases from 30 to 333 °C, and the radiant force of the surface increases to the original by ____ times.
6. There is an air layer with a hot surface temperature of 300 °C and an emissivity of 0.5 and a cold surface of 100 °C and an emissivity of 0.8. When the surface size is much smaller than the thickness of the air layer, the radiation heat transfer between the two surfaces is _____.
7. (Multiple choice) The characteristics of an object that can emit thermal radiation are _____.
   A. The temperature of the object is greater than 0 K
   B. No transmission medium is required
   C. With high temperature
   D. Black surface
8. (Multiple choice) A surface having diffuse radiation property, the surface is _____.
   A. $I_\theta = $ constant
   B. $E_\theta = $ constant
   C. $E = I\pi$
   D. $E_\theta = E_n \cos\theta$

9. According to Planck's law, the relationship between the monochromatic radiant force of a blackbody and its wavelength is a unimodal function, then the corresponding wavelength of the peak value is ____.
   A. Temperature independent
   B. Decreases linearly with increasing temperature
   C. Increases with increasing temperature
   D. Inversely proportional to the absolute temperature
10. The relation between the emissivity of an object and its absorption rate can be determined by Kirchhoff's law and verified by experiments. The following are unconditionally true:
    A. $\varepsilon_T = a_T$
    B. $\varepsilon_{\theta,T} = a_{\theta,T}$
    C. $\varepsilon_{\lambda,T} = a_{\lambda,T}$
    D. $\varepsilon_{\lambda,\theta,T} = a_{\lambda,\theta,T}$
11. The radiant force of the actual object can be expressed as
    A. $E = aE_b$
    B. $E = \dfrac{E_b}{a}$
    C. $E = \varepsilon E_b$
    D. $E = \dfrac{E_b}{\varepsilon}$

Answers

1. $\sigma_b T^4 \pi d^2 / 4$
2. D
3. 148,970 W/m^2
4. 0.76 μm
5. 16
6. 2,228 W/m^2
7. A.B
8. A.B.C
9. D
10. D
11. C

Chapter 13
Heat transfer process and heat exchanger

Main contents of this chapter:

(1) Several common heat transfer processes such as heat conduction through flat wall, tube wall, and fin wall

(2) Basic structure and characteristics of common heat exchanger in industry

(3) Heat transfer calculation method of heat exchanger (key point)

(4) Methods to strengthen or weaken the heat transfer process

13.1 Heat transfer process

13.1.1 Heat transfer through a flat wall

The heat transfer process refers to the process in which heat is transferred from the fluid on one side of the solid wall to the fluid on the other side, and it exists widely in various types of heat exchange equipment. The formula for calculating the heat transfer process through a single-layer flat wall is given in the previous chapter. For the situation shown in Fig. 13.1, the heat flow through the flat wall can be calculated by the following equations:

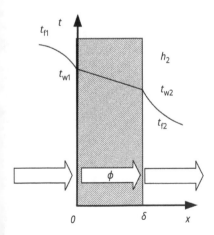

Fig. 13.1: Heat transfer through a flat wall.

$$\Phi = \frac{t_{f1} - t_{f2}}{R_{h1} + R_\lambda + R_{h2}} \tag{13.1}$$

$$\Phi = Ak(t_{f1} - t_{f2}) = Ak\Delta t \tag{13.2}$$

https://doi.org/10.1515/9783111329703-014

Total heat transfer coefficient:

$$k = \frac{1}{\dfrac{1}{h_1} + \dfrac{\delta}{\lambda} + \dfrac{1}{h_2}} \tag{13.3}$$

The thermal resistance network diagram of the heat transfer process through a single-layer flat wall is shown in Fig. 13.2.

t_{f1} R_{h1} t_{w1} R_{λ} t_{w2} R_{h2} t_{f2} Fig. 13.2: Thermal resistance network.

On the basis of the above equations, for the steady-state heat transfer through multilayer flat wall without an internal heat source (Fig. 13.3), using the concept of thermal resistance, the expression equation of heat transfer heat flow can be written as:

$$\Phi = \frac{t_{f1} - t_{f2}}{R_{h1} + \sum_{i=1}^{n} R_{\lambda i} + R_{h2}} = Ak(t_{f1} - t_{f2}) = Ak\Delta t \tag{13.4}$$

Total heat transfer coefficient:

$$k = \frac{1}{\dfrac{1}{h_1} + \sum_{i=1}^{n} \dfrac{\delta_i}{\lambda_i} + \dfrac{1}{h_2}} \tag{13.5}$$

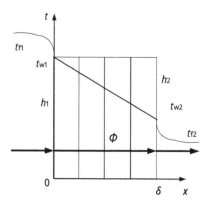

Fig. 13.3: Heat transfer through multilayer flat wall.

13.1.2 Heat transfer through the wall of tube

As shown in Fig. 13.4, a single-layer circular tube with inner and outer radii of r_1 and r_2, and a length of l, has a constant thermal conductivity of λ. The temperature of the fluid inside the tube is t_{f1} and the temperature of the fluid outside the tube is t_{f2}. The temperature of the inner wall of the tube is t_{w1} and the temperature of the outer wall surface is t_{w2}. The convective heat transfer coefficients on both sides are h_1 and h_2, respectively.

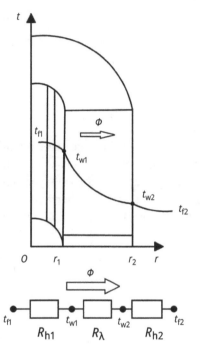

Fig. 13.4: Heat transfer through the wall of tube.

The heat transfer process consists of three parts: convective heat transfer inside the round tube wall, heat conduction in the tube wall, and convective heat transfer outside the round tube wall. The heat flow of these three parts can be expressed as

$$\Phi = \pi d_1 l h_1 (t_{f1} - t_{w1}) = \frac{t_{f1} - t_{w1}}{\dfrac{1}{\pi d_1 l h_1}} = \frac{t_{f1} - t_{w1}}{R_{h_1}} \tag{13.6}$$

$$\Phi = \frac{t_{w1} - t_{w2}}{\dfrac{1}{2\pi \lambda l} \ln \dfrac{d_2}{d_1}} = \frac{t_{w1} - t_{w2}}{R_\lambda} \tag{13.7}$$

$$\Phi = \pi d_2 l h_2 (t_{w2} - t_{f2}) = \frac{t_{w2} - t_{f2}}{\dfrac{1}{\pi d_2 l h_2}} = \frac{t_{w2} - t_{f2}}{R_{h_2}} \tag{13.8}$$

In the case of steady state, the Φ in the above three formulas is the same, so we can get:

$$\Phi = \frac{t_{f1} - t_{f2}}{\dfrac{1}{\pi d_1 l h_1} + \dfrac{1}{2\pi\lambda l}\ln\dfrac{d_2}{d_1} + \dfrac{1}{\pi d_2 l h_2}} = \frac{t_{f1} - t_{f2}}{R_{h_1} + R_\lambda + R_{h_2}} \tag{13.9}$$

The above equation can be written as

$$\Phi = \pi d_2 l k_o (t_{f1} - t_{f2}) = \pi d_2 l k_o \Delta t \tag{13.10}$$

Heat transfer coefficient based on the outer wall area of the tube (overall heat transfer coefficient):

$$k_o = \frac{1}{\dfrac{d_2}{d_1}\dfrac{1}{h_1} + \dfrac{d_2}{2\lambda}\ln\dfrac{d_2}{d_1} + \dfrac{1}{h_2}} \tag{13.11}$$

The heat transfer process from the single-layer circular tube wall can be extended to the heat transfer process of the n-layer circular tube wall. For the steady-state heat transfer in an n-layer tube, assuming that the thermal conductivity of each layer material is constant, there is no contact thermal resistance, and the total thermal resistance is the sum of the thermal resistances connected in series. The heat flow can be written as

$$\Phi = \frac{t_{f1} - t_{f2}}{R_{h_1} + R_\lambda + R_{h_2}} \tag{13.12}$$

$$\Phi = \frac{t_{f1} - t_{f2}}{R_{h_1} + \sum_{i=1}^{n} R_{\lambda_i} + R_{h_2}}$$

$$= \frac{t_{f1} - t_{f2}}{\dfrac{1}{\pi d_1 l h_1} + \sum_{i=1}^{n} \dfrac{1}{2\pi\lambda_i l}\ln\dfrac{d_{i+1}}{d_i} + \dfrac{1}{\pi d_{n+1} l h_2}} \tag{13.13}$$

13.1.3 Heat transfer process through fin wall

For the heat transfer process with large difference of heat transfer coefficient on both sides, adding fins on the side wall with large heat transfer coefficient (expand the heat exchange area) is an effective measure to enhance or weaken heat transfer. The following is an example of the heat transfer process through a flat wall.

As shown in Fig. 13.5, assuming that $h_1 \gg h_2$, fins can be added to the right-side wall where convection heat transfer is weak in order to enhance heat transfer. Assuming that the thermal conductivity of the material and the surface convective heat transfer coefficient are both constant, in the steady state, the heat flow is written for the three parts of the heat transfer process:

$$\Phi = A_1 h_1 (t_{f1} - t_{w1}) = \frac{t_{f1} - t_{w1}}{\dfrac{1}{A_1 h_1}} \tag{13.14}$$

$$\Phi = \frac{t_{w1} - t_{w2}}{\dfrac{\delta}{A_1 \lambda}} \tag{13.15}$$

$$\Phi = A_2' h_2 \left(t_{w2}' - t_{f2} \right) + A_2'' h_2 \left(t_{w2}'' - t_{f2} \right) \tag{13.16}$$

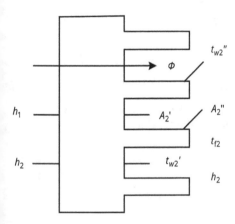

Fig. 13.5: Heat transfer process through fin wall.

According to the definition of fin efficiency,

$$\eta_f = \frac{A_2'' h_2 \left(t_{w2}'' - t_{f2} \right)}{A_2'' h_2 \left(t_{w2}' - t_{f2} \right)} = \frac{t_{w2}'' - t_{f2}}{t_{w2}' - t_{f2}} \tag{13.17}$$

$$\Phi = \left(A_2' + A_2'' \eta_f \right) h_2 \left(t_{w2}' - t_{f2} \right)$$

$$= A_2 \eta h_2 \left(t_{w2}' - t_{f2} \right) = \frac{t_{w2}' - t_{f2}}{\dfrac{1}{A_2 \eta h_2}} \tag{13.18}$$

$$A_2 = A_2' + A_2'' \tag{13.19}$$

Total fin efficiency:

$$\eta = \frac{\left(A_2' + A_2'' \eta_f \right)}{A_2} \tag{13.20}$$

In general, $A_2'' \gg A_2'$, $A_2 \approx A_2''$, $\eta \approx \eta_f$.

All the above are available:

$$\Phi = \frac{t_{f1} - t_{f2}}{\dfrac{1}{A_1 h_1} + \dfrac{\delta}{A_1 \lambda} + \dfrac{1}{A_2 \eta h_2}} = A_1 k_1 (t_{f1} - t_{f2}) = A_1 k_1 \Delta t \tag{13.21}$$

k_1 is called the heat transfer coefficient based on the surface area of the smooth wall, and the expression is

$$k_1 = \frac{1}{\dfrac{1}{h_1} + \dfrac{\delta}{\lambda} + \dfrac{1}{\beta \eta h_2}} \tag{13.22}$$

Finned coefficient:

$$\beta = \frac{A_2}{A_1} \tag{13.23}$$

Compare the convection heat transfer resistance on the fin side:
Before finned:

$$R_{h_2} = \frac{1}{A_1 h_2} \tag{13.24}$$

After finned:

$$R'_{h_2} = \frac{1}{A_2 \eta h_2} \tag{13.25}$$

The ratio of heat resistance before and after finned is

$$R_{h_2} / R'_{h_2} = \beta \eta \tag{13.26}$$

From the definition of finned coefficient, its size depends on fin height and fin spacing. Generally, the fin spacing should be greater than 2 times the maximum thickness of the boundary layer. The finned coefficient should be chosen reasonably. In engineering, when $h_1/h_2 = 3$–5, generally choose low fin with smaller β; when $h_1/h_2 > 10$, generally choose high fin with larger β. To effectively enhance heat transfer, fins should always be added on the side with the smaller heat transfer coefficient.

In engineering, the heat transfer coefficient k_2 based on the fin surface area A_2 is usually used to calculate:

$$\Phi = A_2 k_2 \Delta t \tag{13.27}$$

$$k_2 = \frac{1}{\dfrac{1}{h_1} \beta + \dfrac{\delta}{\lambda} \beta + \dfrac{1}{\eta h_2}} \tag{13.28}$$

13.1.4 Combined heat transfer

Generally, this refers to the heat transfer process in which convection and radiation exist simultaneously.

In engineering, the radiant heat transfer is generally converted into convective heat transfer for calculation, so the surface heat transfer coefficient of radiant heat transfer is introduced:

$$h_r = \frac{\Phi_r}{A(t_w - t_f)} \tag{13.29}$$

Φ_r is the radiation heat exchange. The composite surface heat transfer coefficient is the sum of the convective surface heat transfer coefficient and the radiation surface heat transfer coefficient. The total heat exchange is the sum of convective heat exchange and radiation heat exchange:

$$h = h_c + h_r \tag{13.30}$$

$$\Phi = \Phi_c + \Phi_r = (h_c + h_r)A(t_w - t_f) = hA(t_w - t_f) \tag{13.31}$$

13.2 Heat exchanger

Heat exchanger is the equipment for heat transfer from hot fluid to cold fluid. A heat exchanger allows heat to be transferred between two fluids without the fluids contacting each other.

Heat exchangers are widely used in industry and daily life. In the 1960s, the first domestically produced shell-and-tube heat exchanger, the first plate heat exchanger, and the first spiral plate heat exchanger appeared. After the 1980s, a new trend of self-developed heat transfer technology appeared in China, and a large number of enhanced heat transfer elements were introduced to the market: heater, high-efficiency reboiler, high-efficiency condenser, double-shell heat exchanger, plate-and-shell heat exchanger, surface evaporative air cooler, and a number of excellent high-efficiency heat exchangers. After the twenty-first century, with the rapid development of modern industry, energy-centered environmental and ecological problems have become increasingly intensified. While all countries in the world are looking for new energy sources, they also pay more attention to the research and development of new ways to save energy. A large number of enhanced heat transfer technologies have been applied, and the heat exchanger industry has ushered in a big leap in the technical level, and began to emphasize the optimal matching of comprehensive cost and cooling effect. In the new era, large-scale, high-efficiency, and energy-saving products have become the main trends. The application of enhanced heat transfer technology can not only save energy and protect the environment but also greatly save investment costs. The market size of heat exchangers continues to expand due to their wide application in industrial sectors such as chemical, petroleum, power, and atomic energy. According to statistics, the market size of China's heat exchanger industry has increased from 116.8 billion yuan in 2019 to 142.6 billion yuan in 2021, with an average annual compound growth rate of 10.5%. It is expected that the market for heat exchangers will further expand in the future.

13.2.1 Classification of heat exchangers

There are many types of heat exchangers. According to the working principle of heat exchanger, it can be divided into:

(1) Hybrid: The cold and hot fluids in the heat exchanger contact directly and mix with each other to realize heat exchange. This is also known as direct contact heat exchanger. The application of such heat exchangers is limited when the cold and hot fluids cannot be directly mixed.

(2) Regenerative: Both cold and hot fluids flow through the same heat exchange surface (regenerator) of the heat exchanger in turn to achieve unsteady heat exchange. Specifically, it transfers heat from a high-temperature fluid to a low-temperature fluid through a regenerator composed of solid materials. The heat medium first reaches a certain temperature by heating the solid material, and then the cold medium is heated through the solid material to achieve the heat transfer. In this heat exchanger, the non-steady heat transfer process is carried out. Regenerative heat exchangers include rotary and valve switching.

Figure 13.6 shows the schematic diagram of the regenerative high-temperature air combustion technology. It applies the idea of thermal storage. The normal temperature air enters the regenerator through the reversing valve, and is heated to the temperature closed to the furnace in a very short time. After the high-temperature hot air enters the furnace, the fuel is burned in a state of lean oxygen. At the same time, the high-temperature flue gas in the furnace enters another regenerator to store the sensible heat in the regenerator, and then the low-temperature flue gas is discharged into the atmosphere through the reversing valve. The reversing valve switches at a certain frequency so that the two regenerators are in the alternate working state of heat storage and heat release. The regenerative high-temperature air combustion technology is suitable for many fields. It can recover most of the sensible heat of the flue gas, make the temperature field in the furnace relatively uniform, and effectively reduce the generation of high-temperature thermal NO_x.

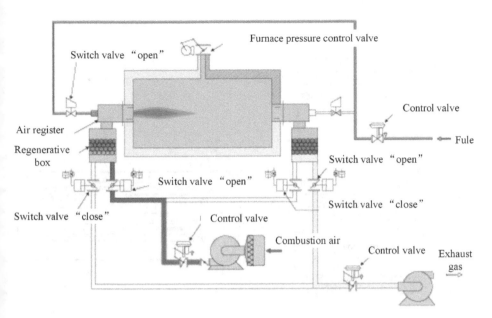

Fig. 13.6: Regenerative high-temperature air combustion technology.

(3) Dividing wall type: The cold and hot fluids in the heat exchanger are separated by the wall, and the heat transfer from the hot fluid to the cold fluid consists of three parts: the convective heat transfer between the hot fluid and the wall, the heat conduction of the wall, and the convective heat transfer between the wall and the cold fluid.

The partition heat exchanger is the most widely used heat exchanger. The structure of several specific partition heat exchangers will be introduced below.

According to the structure, the dividing wall-type heat exchanger can be divided into: shell-and-tube heat exchanger, finned tube heat exchanger, plate heat exchanger, plate-and-fin heat exchanger, and spiral-plate heat exchanger.

(1) Shell-and-tube heat exchanger (also known as double-pipe heat exchanger):
This heat exchanger accounts for 60% of heat exchangers in use today. It can handle large flows, low temperatures and pressures, and high temperatures and pressures. Heat exchanger consists of tube and shell. Cold and hot fluids flow through the inner tube and the interlayer, respectively. According to the relative flow direction of hot and cold fluids, it can be divided into parallel flow and counter flow (Fig. 13.7). The heat exchange area of the shell-and-tube heat exchanger (Fig. 13.8) is often small, which is suitable for the situation when the heat transfer is not large or the fluid flow is small.

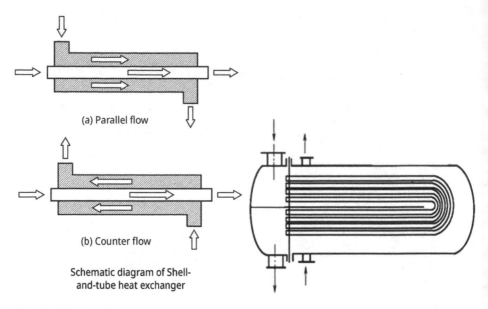

(a) Parallel flow

(b) Counter flow

Schematic diagram of Shell-
and-tube heat exchanger

Fig. 13.7: Schematic diagram.

Fig. 13.8: Shell-and-tube heat exchanger.

(2) Finned tube heat exchanger:

The heat exchanger composed of finned tube bundles is suitable for the heat transfer between the liquid inside the tube and the gas outside the tube, and the heat transfer coefficients on the two sides of the surface differ greatly, such as car water tank radiator and air conditioner evaporator and condenser. As shown in Fig. 13.9.

Since the fins are often added on the air side, the heat exchange area on the air side is greatly increased, thereby improving the heat transfer coefficient. The shape and structure of the fins are important considerations when designing this heat exchanger.

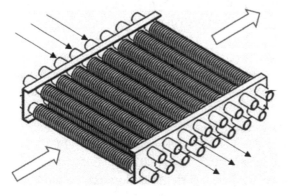

Fig. 13.9: Finned tube heat exchanger.

(3) Plate heat exchanger

It is composed of several corrugated metal plates, as shown in Fig. 13.10. The four corners of the metal plate are provided with corner holes, and adjacent plates are separated. The cold and hot fluids flow through a corner hole, respectively, and flow along the flow channel set by the corrugations at intervals, exchange heat during the flow, and finally flow out from the diagonal corner holes, respectively. There are a lot of advantages of plate heat exchanger, such as high heat transfer coefficient, relatively small resistance, compact structure, and convenient disassembly and cleaning.

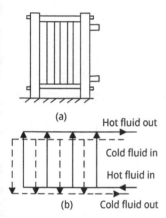

(a)

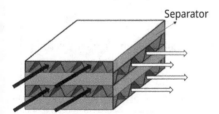

Hot fluid out

Cold fluid in

Hot fluid in

(b) Cold fluid out

Fig. 13.10: Plate heat exchanger.

(4) Plate-and-fin heat exchanger:

It is made of metal plate and corrugated plate fin by lamination and staggered welding with compact structure, as shown in Fig. 13.11. The disadvantage is that it is difficult to clean and repair. It is suitable for heat transfer between clean and non-corrosive fluids.

Separator

Fig. 13.11: Plate-and-fin heat exchanger.

(5) Spiral-plate heat exchanger:

It is made of two parallel metal plates, the metal plate forms two spiral channels as shown in Fig. 13.12, in which the cold and hot fluids flow, respectively. The advantages are simple technology, low price, and small flow resistance. The disadvantages are not easy to clean and low-pressure bearing capacity.

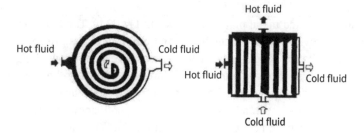

Fig. 13.12: Spiral-plate heat exchanger.

The different flow modes of the heat exchangers are shown in Fig. 13.13, such as parallel flow, counter flow, cross flow, and mixed flow. In the case of the same inlet temperature, flow rate, and heat exchange surface area of cold and hot fluid, the flow pattern affects the outlet temperature, heat exchange temperature difference, heat exchange, and the temperature distribution in the heat exchanger.

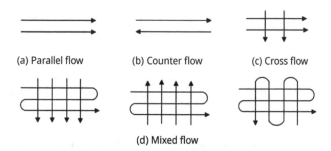

Fig. 13.13: Different flow modes.

13.2.2 Heat transfer calculation of heat exchangers

According to the different purpose, the heat transfer calculation of heat exchangers can be divided into two types:

Design calculation: According to the conditions and requirements of heat exchange, a new heat exchanger is designed, so the type, structure, and heat exchange area of the heat exchanger need to be determined.

Checking calculation: To check whether the existing heat exchanger can meet the heat exchange requirements, it is generally necessary to calculate the outlet temperature, amount of heat exchange, and flow resistance of the fluid.

The common calculation method of heat exchanger is **mean temperature difference method**.

(1) Mean temperature difference of heat transfer in heat exchanger

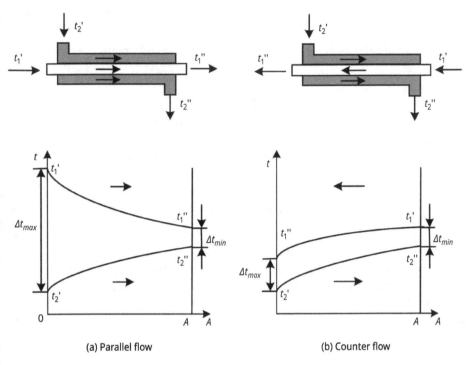

(a) Parallel flow   **(b) Counter flow**

Fig. 13.14: Schematic diagram of fluid temperature change along the process: 1, hot fluid; 2, cold fluid; ', inlet; ", outlet.

In the heat exchanger, because the cold and hot fluids continuously exchange heat along the heat exchange surface, their temperature difference is not constant as shown in Fig. 13.14. In the heat transfer calculation of the heat exchanger, the temperature difference term should take the average temperature difference Δt_m of the entire heat transfer surface of the heat exchanger. Heat transfer process of the heat exchanger:

$$\Phi = kA\Delta t_m \tag{13.32}$$

The heat transfer temperature difference between the inlet and outlet of the heat exchanger is $\Delta t'$ and $\Delta t''$. If use Δt_{max} for the larger of the two, Δt_{min} for the smaller, then the average temperature difference of the heat exchanger can be calculated by the following equation. Δt_m is also called log mean temperature difference:

$$\Delta t_m = \frac{\Delta t_{max} - \Delta t_{min}}{\ln \dfrac{\Delta t_{max}}{\Delta t_{min}}} \tag{13.33}$$

In engineering, when

$$\frac{\Delta t_{max}}{\Delta t_{min}} \leq 2 \tag{13.34}$$

Δt_m can be expressed as the arithmetic mean temperature difference:

$$\Delta t_m \approx \frac{\Delta t_{max} + \Delta t_{min}}{2} \tag{13.35}$$

(2) Mean temperature difference method for heat transfer calculation of heat exchanger:

Basic equations for heat transfer calculation of heat exchanger:

$$\Phi = kA\Delta t_m \tag{13.32}$$

$$\Phi = q_{m1}c_{p1}\left(t_1' - t_1''\right) \tag{13.35}$$

$$\Phi = q_{m2}c_{p2}\left(t_2'' - t_2'\right) \tag{13.36}$$

q_{m1} and q_{m2} are the mass flow rates of the hot and cold fluids, respectively, c_{p1} and c_{p2} are the constant pressure specific heat capacities of the cold and hot fluids, respectively. If c_{p1} and c_{p2} are known, the three equations have eight independent variables (Φ, k, A, q_{m1}, q_{m2}, and three of the four t). As long as we know five of them, we can find the other three variables.

1) Design calculation:

According to the requirements of the production task, given the mass flow of cold and hot fluids q_{m1}, q_{m2} and three of the four inlet and outlet temperatures, it is necessary to determine the type and structure of the heat exchanger. Calculate the heat transfer coefficient k and the heat exchange area A.

Design calculation steps:

a) According to the given heat transfer conditions, such as fluid properties, temperature, and pressure range, select the type of heat exchanger, arrange the heat transfer surface, and calculate the surface heat transfer coefficients h_1, h_2, and k of convective heat transfer on both sides.

b) According to the given conditions, eqs. (13.36) and (13.37), find out the unknown inlet and outlet temperature, and the heat exchange amount Φ.

c) Calculate the average temperature difference Δt_m from four inlet and outlet temperatures and flow patterns.

d) Calculate the required heat exchange area A from eq. (13.32).

e) Calculate the flow resistance of the heat exchanger. If the resistance is too large, the investment and operation cost of the equipment will be increased. The scheme must be changed and redesigned.

2) Checking calculation:

Generally, five parameters (heat exchange area A, mass flow rate of fluid on both sides q_{m1}, q_{m2}, and inlet temperatures (t'_1, t'_2)) are known. It is necessary to calculate the outlet temperature of hot and cold fluids t'_1, t'_2 and heat exchange amount Φ.

Because the outlet temperature of the fluid on both sides is unknown, the mean temperature difference of heat transfer cannot be calculated, the physical properties of the fluid cannot be determined, and h and k cannot be calculated, the remaining unknown quantities cannot be calculated directly by eqs. (13.32), (13.36), and (13.37). In this case, the trial method is usually used.

Calculation steps:

(a) First, assume the outlet temperature of one fluid and use the heat balance equation to calculate the heat exchange amount Φ' and the outlet temperature of the other fluid.

(b) Calculate the mean temperature difference Δt_m according to the four inlet and outlet temperatures of the fluid.

(c) Calculate the surface heat transfer coefficients h_1 and h_2 on both sides of the heat exchange surface, and then the heat transfer coefficient k is obtained.

(d) Calculate the heat exchange amount Φ'' from eq. (13.32).

(e) Compare Φ' and Φ'' if the difference is large, assume the outlet temperature of the fluid again, and repeat the above calculation until it is satisfied.

Example

There is a double-pipe heat exchanger (Fig. 13.15). The flow rate of hot fluid is $q_{m1} = 0.1$ kg/s, the specific constant pressure heat capacity is $c_{p1} = 2{,}100$ J/(kg·K), the inlet temperature $t'_1 = 220$ °C, the flow rate of cold fluid is $q_{m2} = 0.2$ kg/s, the specific constant pressure heat capacity is $c_{p2} = 4{,}200$ J/(kg·K), the inlet temperature $t'_2 = 20$ °C, and the outlet temperature $t''_2 = 38$ °C. The heat transfer coefficient of the heat exchanger is $k = 400$ W/(m²·K). Try to find out the required heat exchange area when the cold and hot fluids flow in parallel:

$$\Phi = kA\Delta t_m$$

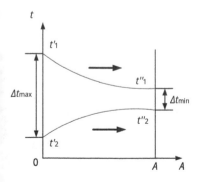

Fig. 13.15: Temperature change in heat exchanger.

13.3 Enhancing and weakening of heat transfer

Heat transfer engineering technology is a science and engineering technology developed according to the needs of modern industrial production and scientific practice. Its main task is to control and optimize the heat transfer process according to the requirements of industrial production and scientific practice. The application example is shown in Fig. 13.16.

Depending on the application purpose, there are two directions of heat transfer engineering technology:
(1) Enhancing heat transfer technology
(2) Heat insulation technology

Fig. 13.16: Examples in daily life.

Main purpose of enhancing heat transfer:
(1) Increase the amount of heat transfer Φ
(2) Reduce heat transfer area A, reduce equipment size, and reduce material consumption
(3) Reduce the temperature T of high-temperature components to ensure the safe operation of the equipment
(4) Reduce the transmission power of heat-carrying fluid q_m.

Main purpose of weakening heat transfer:
(1) Reduce heat loss and save energy.
(2) Maintain the artificial low-temperature environment and reduce the external heat input.
(3) Protect personal safety from heat or cold, and create a working and living environment with appropriate temperature.

Whether it is to strengthen or weaken heat transfer, we usually start from two aspects: changing heat transfer temperature difference and changing heat transfer resistance.

Take the heat transfer process in heat exchanger for example:

$$\Phi = kA\Delta t_{\mathrm{m}} = \frac{\Delta t_{\mathrm{m}}}{\dfrac{1}{kA}} = \frac{\Delta t_{\mathrm{m}}}{R_{\mathrm{k}}} = \frac{\Delta t_{\mathrm{m}}}{R_{\mathrm{h1}} + R_{\lambda} + R_{\mathrm{h2}}} \tag{13.37}$$

From the equation above, there are two ways to enhance heat transfer:

(1) Increase heat transfer temperature difference
In the case of the same inlet and outlet temperatures of cold and hot fluids, the mean temperature difference of counter flow is the largest and that of parallel flow is the smallest. Therefore, from the perspective of heat transfer enhancement, the heat exchanger should be arranged as counter flow as possible. However, in most practical situations, the heat transfer temperature difference is often limited by the objective environment, production process, and equipment conditions, so there is often not much room for consideration to improve the heat transfer effect by increasing the heat transfer temperature difference.

(2) Reduce thermal resistance R_{k}

$$R_{\mathrm{k}} = R_{\mathrm{h1}} + R_{\lambda} + R_{\mathrm{h2}} = \frac{1}{Ah_1} + \frac{\delta}{A\lambda} + \frac{1}{Ah_2} \tag{13.38}$$

It can be seen from the calculation equation of the total thermal resistance that the size of the heat transfer thermal resistance is related to the heat transfer area. Arrange more heat transfer surfaces to increase the total heat transfer area A, which can reduce the total heat transfer resistance and increase the heat transfer.

Without considering radiation heat transfer, the total thermal resistance consists of one conduction thermal resistance and two convection thermal resistances. In the case of dirt adhesion on the surface, the dirt thermal resistance also needs to be considered. In engineering, the heat exchange surface of most heat exchange equipment is made of metal, and its conductive thermal resistance is very small compared with the convection thermal resistance, so reducing the total thermal resistance should start with reducing the fouling thermal resistance and the two convective thermal resistance terms R_{h1} and R_{h2}.

If there is a large difference between the two thermal resistances, we should grasp the main contradiction and try to reduce the maximum thermal resistance.

Methods to enhance convective heat transfer are:
1) Extended heat exchange surface (with fins)

2) Change the shape, size, and position of the heat exchange surface
For example, small diameter tube or elliptical tube instead of a circular tube (reducing the equivalent diameter) used in forced convective turbulent heat transfer in the tube can achieve the effect of enhancing convective heat transfer.

In addition, in the case of natural convection and condensation heat transfer outside the tube, the surface heat transfer coefficient of horizontal tube is higher than that of vertical tube.

3) Change the surface condition of heat exchange surface
By increasing the surface roughness, the turbulent heat transfer of single-phase fluid can be enhanced. The formation of a porous layer on the heat transfer surface can enhance the boiling heat transfer. It is a practical technique to enhance the condensation heat transfer by machining groove or thread on the heat exchange surface or treating the heat exchange surface to cause pearl-shaped condensation (Fig. 13.17). Another example is, changing the surface blackness can enhance radiation heat transfer.

  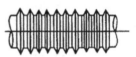

Fig. 13.17: Some ways to enhance heat transfer.

4) Change the flow condition of fluid
The intensity of turbulent heat transfer is greater than that of laminar flow, and the heat resistance of convective heat transfer is mainly concentrated in the boundary layer. The main methods to enhance convective heat transfer are to increase the flow velocity to realize turbulent heat transfer, enhance fluid disturbance, destroy the boundary layer, and laminar bottom layer.

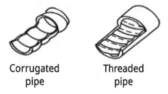

Corrugated pipe   Threaded pipe

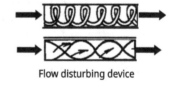

Flow disturbing device

Fig. 13.18: Corrugated and threaded pipe.   Fig. 13.19: Flow disturbing device.

For example, the heat exchange surface is processed into corrugation (Fig. 13.18), and the flow disturbing device is added to the flow channel (Fig. 13.19). By vibrating the heat exchange surface or oscillating the fluid, the fluid disturbance is enhanced, the boundary layer is destroyed, and the purpose of strengthening convection heat transfer is achieved.

13.4 Summary

Focus on the following:
(1) Analysis method of heat transfer process (heat transfer process of single layer, multilayer flat wall, cylinder wall, and fin wall)
(2) Mean temperature difference method for heat exchanger calculation
(3) The principle and method of enhancing and weakening heat transfer.

Exercises

1. There is a counter flow oil–water heat exchanger. The inlet temperature of the oil is $t_1' = 100$ °C, the outlet temperature is $t_1'' = 50$ °C, the density of the oil is $\rho = 860$ kg/m³, and the specific heat capacity is $c_p = 2.1$ kJ/kg · K. The inlet temperature of cooling water is $t_2' = 20$ °C, the outlet temperature is $t_2'' = 45$ °C, and the flow rate $q_{m2} = 3$ kg/s. Heat transfer coefficient of heat exchanger $k = 300$ W/(m² · K). Try to find out the:
    (1) heat transfer amount of heat exchanger;
    (2) flow rate of oil; and
    (3) heat transfer area of heat exchanger.

Answers

1. (1) $\quad c_{p2} = 4.174$ kJ/(kg K)

    $$\Phi = q_{m2}c_{p2}\left(t_2'' - t_2'\right) = 3 \times 4.174 \times 25 = 313.05 \text{ kW}$$

    (2) $\quad \Phi = q_{m1}c_{p1}\left(t_1' - t_1''\right)$

    $$\Downarrow$$

    $$q_{m1} = \frac{\Phi}{c_{p1}\left(t_1' - t_1''\right)} = \frac{313.05}{2.1 \times 50} = 2.98 \text{ kg/s}$$

    (3) $\quad \Delta t' = t_1' - t_2'' = 100 - 45 = 55\,°C$

    $$\Delta t'' = t_1'' - t_2' = 50 - 20 = 30\,°C$$

    $$\Delta t_m = \frac{\Delta t' - \Delta t''}{\ln\dfrac{\Delta t'}{\Delta t''}} = 41.24\,°C$$

    $$\Phi = kA\Delta t_m \rightarrow A = \frac{\Phi}{k\Delta t_m} = 25.3 \text{ m}^2$$

Appendix
Table A1: Saturated and unsaturated vapor (by temperature)

Tempera-ture	Pressure	Specific volume		Specific enthalpy		Latent heat of vapori-zation	Specific entropy	
		Liquid	Vapor	Liquid	Vapor		Liquid	Vapor
t (°C)	p (MPa)	m³/kg	m³/kg	kJ/kg	kJ/kg	r (kJ/kg)	kJ/(kg · K)	kJ/(kg · K)
0.00	0.0006112	0.00100022	206.154	−0.05	2,500.51	2,500.6	−0.0002	9.1544
0.01	0.0006117	0.00100021	206.012	0.01[①]	2,500.53	2,500.5	0.0000	9.1541
1	0.0006571	0.00100018	192.464	4.18	2,502.35	2,498.2	0.0153	9.1278
2	0.0007059	0.00100013	179.787	8.39	2,504.19	2,495.8	0.0306	9.1014
3	0.0007580	0.00100009	168.041	12.61	2,506.03	2,493.4	0.0459	9.0752
4	0.0008135	0.00100008	157.151	16.82	2,507.87	2,491.1	0.0611	9.0493
5	0.0008725	0.00100008	147.048	21.02	2,509.71	2,488.7	0.0763	9.0236
6	0.0009352	0.00100010	137.670	25.22	2,511.55	2,486.3	0.0913	8.9982
7	0.0010019	0.00100014	128.961	29.42	2,513.39	2,484.0	0.1063	8.9730
8	0.0010728	0.00100019	120.868	33.62	2,515.23	2,481.6	0.1213	8.9480
9	0.0011480	0.00100026	113.342	37.81	2,517.06	2,479.3	0.1362	8.9233
10	0.0012279	0.00100034	106.341	42.00	2,518.90	2,476.9	0.1510	8.8988
11	0.0013126	0.00100043	99.825	46.19	2,520.74	2,474.5	0.1658	8.8745
12	0.0014025	0.00100054	93.756	50.38	2,522.57	2,472.2	0.1805	8.8504
13	0.0014977	0.00100066	88.101	54.57	2,524.41	2,469.8	0.1952	8.8265
14	0.0015985	0.00100080	82.828	58.76	2,526.24	2,467.5	0.2098	8.8029
15	0.0017053	0.00100094	77.910	62.95	2528.07	2465.1	0.2243	8.7794
16	0.0018183	0.00100110	73.320	67.13	2529.90	2462.8	0.2388	8.7562
17	0.0019377	0.00100127	69.034	71.32	2,531.72	2,460.4	0.2533	8.7331
18	0.0020640	0.00100145	65.029	75.50	2,533.55	2,458.1	0.2677	8.7103
19	0.0021975	0.00100165	61.287	79.68	2,535.37	2,455.7	0.2820	8.6877
20	0.0023385	0.00100185	57.786	83.86	2,537.20	2,453.3	0.2963	8.6652
22	0.0026444	0.00100229	51.445	92.23	2,540.84	2,448.6	0.3247	8.6210
24	0.0029846	0.00100276	45.884	100.59	2,544.47	2,443.9	0.3530	8.5774
26	0.0033625	0.00100328	40.997	108.95	2,548.10	2,439.2	0.3810	8.5347
28	0.0037814	0.00100383	36.694	117.32	2,551.73	2,434.4	0.4089	8.4927
30	0.0042451	0.00100442	32.899	125.68	2,555.35	2,429.7	0.4366	8.4514
35	0.0056263	0.00100605	25.222	146.59	2,564.38	2,417.8	0.5050	8.3511
40	0.0073811	0.00100789	19.529	167.50	2,573.36	2,405.9	0.5723	8.2551
45	0.0095897	0.00100993	15.2636	188.42	2,582.30	2,393.9	0.6386	8.1630
50	0.0123446	0.00101216	12.0365	209.33	2,591.19	2,381.9	0.7038	8.0745
55	0.015752	0.00101455	9.5723	230.24	2,600.02	2,369.8	0.7680	7.9896
60	0.019933	0.00101713	7.6740	251.15	2,608.79	2,357.6	0.8312	7.9080
65	0.025024	0.00101986	6.1992	272.08	2,617.48	2,345.4	0.8935	7.8295
70	0.031178	0.00102276	5.0443	293.01	2,626.10	2,333.1	0.9550	7.7540

https://doi.org/10.1515/9783111329703-015

(continued)

Tempera-ture	Pressure	Specific volume		Specific enthalpy		Latent heat of vapori-zation	Specific entropy	
t (°C)	p (MPa)	Liquid	Vapor	Liquid	Vapor	r (kJ/kg)	Liquid	Vapor
		m³/kg	m³/kg	kJ/kg	kJ/kg		kJ/ (kg · K)	kJ/ (kg · K)
26	0.0033625	0.00100328	40.997	108.95	2,548.10	2,439.2	0.3810	8.5347
28	0.0037814	0.00100383	36.694	117.32	2,551.73	2,434.4	0.4089	8.4927
30	0.0042451	0.00100442	32.899	125.68	2,555.35	2,429.7	0.4366	8.4514
35	0.0056263	0.00100605	25.222	146.59	2,564.38	2,417.8	0.5050	8.3511
40	0.0073811	0.00100789	19.529	167.50	2,573.36	2,405.9	0.5723	8.2551
45	0.0095897	0.00100993	15.2636	188.42	2,582.30	2,393.9	0.6386	8.1630
50	0.0123446	0.00101216	12.0365	209.33	2,591.19	2,381.9	0.7038	8.0745
55	0.015752	0.00101455	9.5723	230.24	2,600.02	2,369.8	0.7680	7.9896
60	0.019933	0.00101713	7.6740	251.15	2,608.79	2,357.6	0.8312	7.9080
65	0.025024	0.00101986	6.1992	272.08	2,617.48	2,345.4	0.8935	7.8295
70	0.031178	0.00102276	5.0443	293.01	2,626.10	2,333.1	0.9550	7.7540
75	0.038565	0.00102582	4.1330	313.96	2,634.63	2,320.7	1.0156	7.6812
80	0.047376	0.00102903	3.4086	334.93	2,643.06	2,308.1	1.0753	7.6112
85	0.057818	0.00103240	2.8288	355.92	2,651.40	2,295.5	1.1343	7.5436
90	0.070121	0.00103593	2.3616	376.94	2,659.63	2,282.7	1.1926	7.4783
95	0.084533	0.00103961	1.9827	397.98	2,667.73	2,269.7	1.2501	7.4154
100	0.101325	0.00104344	1.6736	419.06	2,675.71	2,256.6	1.3069	7.3545
110	0.143243	0.00105156	1.2106	461.33	2,691.26	2,229.9	1.4186	7.2386
120	0.198483	0.00106031	0.89219	503.76	2,706.18	2,202.4	1.5277	7.1297
130	0.270018	0.00106968	0.66873	546.38	2,720.39	2,174.0	1.6346	7.0272
140	0.361190	0.00107972	0.50900	589.21	2,733.81	2,144.6	1.7393	6.9302
150	0.47571	0.00109046	0.39286	632.28	2,746.35	2,114.1	1.8420	6.8381
160	0.61766	0.00110193	0.30709	675.62	2,757.92	2,082.3	1.9429	6.7502
170	0.79147	0.00111420	0.24283	719.25	2,768.42	2,049.2	2.0420	6.6661
180	1.00193	0.00112732	0.19403	763.22	2,777.74	2,014.5	2.1396	6.5852
190	1.25417	0.00114136	0.15650	807.56	2,785.80	1,978.2	2.2358	6.5071
200	1.55366	0.00115641	0.12732	852.34	2,792.47	1,940.1	2.3307	6.4312
210	1.90617	0.00117258	0.10438	897.62	2,797.65	1,900.0	2.4245	6.3571
220	2.31783	0.00119000	0.086157	943.46	2,801.20	1,857.7	2.5175	6.2846
230	2.79505	0.00120882	0.071553	989.95	2,803.00	1,813.0	2.6096	6.2130
240	3.34459	0.00122922	0.059743	1,037.2	2,802.88	1,765.7	2.7013	6.1422
250	3.97351	0.00125145	0.050112	1,085.3	2,800.66	1,715.4	2.7926	6.0716
260	4.68923	0.00127579	0.042195	1,134.3	2,796.14	1,661.8	2.8850	6.0007
270	5.49956	0.00130262	0.035637	1,184.5	2,789.05	1,604.5	2.8837	5.9292
280	6.41273	0.00133242	0.030165	1,236.0	2,779.08	1,543.1	2.9751	5.8564
290	7.43746	0.00136582	0.025565	1,289.1	2,765.81	1,476.7	3.1594	5.7871
300	8.58308	0.00140369	0.021669	1,344.0	2,748.71	1,404.7	3.2533	5.7042
310	9.85970	0.00144728	0.018343	1,401.2	2,727.01	1,325.9	3.3490	5.6226

(continued)

Tempera-ture	Pressure	Specific volume		Specific enthalpy		Latent heat of vapori-zation	Specific entropy	
		Liquid	Vapor	Liquid	Vapor	r (kJ/kg)	Liquid	Vapor
t (°C)	p (MPa)	m³/kg	m³/kg	kJ/kg	kJ/kg		kJ/(kg · K)	kJ/(kg · K)
320	11.278	0.00149844	0.015479	1,461.2	2,699.72	1,238.5	3.4475	5.5356
330	12.851	0.00156008	0.012987	1,524.9	2,665.30	1,140.4	3.5500	5.4408
340	14.593	0.00163728	0.010790	1,593.7	2,621.32	1,027.6	3.6586	5.3345
350	16.521	0.00174008	0.008812	1,670.3	2,563.39	893.0	3.7773	5.2104
360	18.657	0.00189423	0.006958	1,761.1	2,481.68	720.6	3.9155	5.0536
370	21.033	0.00221480	0.004982	1,891.7	2,338.79	447.1	4.1125	4.8076
371	21.286	0.00227969	0.004735	1,911.8	2,314.11	402.3	4.1429	4.7674
372	21.542	0.00236430	0.004451	1,936.1	2,282.99	346.9	4.1796	4.7173
373	21.802	0.00249600	0.004087	1,968.8	2,237.98	269.2	4.2292	4.6458
373.99	22.064	0.00310600	0.003106	2,085.9	2,085.9	0.00	4.40952	4.4092

374.12: The data in this row are the parameter values of the critical state.

①The exact value is 0.000612 kJ/kg.

Source of data: Tables A1 and A2 are generated using the Engineering Equation Solver (EES) software developed by S. A. Klein and F. L. Alvarado. The routine used in calculations is the highly accurate Steam_IAPWS, which incorporates the 1995 Formulation for the Thermodynamic Properties of Ordinary Water Substance for General and Scientific Use, issued by The International Association for the Properties of Water and Steam (IAPWS). This formulation replaces the 1984 formulation of Haar, Gallagher, and Kell (NBS/NRC Steam Tables, Hemisphere Publishing Co., 1984), which is also available in EES as the routine STEAM. The new formulation is based on the correlations of Saul and Wagner (J. Phys. Chem. Ref. Data, 16, 893, 1987) with modifications to adjust to the International Temperature Scale of 1990. The modifications are described by Wagner and Pruss (J. Phys. Chem. Ref. Data, 22, 783, 1993).

Appendix
Table A2: Saturated and unsaturated vapor (by pressure)

(The pressure in the table refers to the absolute pressure, which should be added after 0.1 MPa.)

Pressure	Temperature	Specific volume		Specific enthalpy		Latent heat of vaporization	Specific entropy	
p (MPa)	t (°C)	Liquid	Vapor	Liquid	Vapor	r (kJ/kg)	Liquid	Vapor
		m³/kg	m³/kg	kJ/kg	kJ/kg		kJ/ (kg · K)	kJ/ (kg · K)
0.0010	6.9491	0.0010001	129.185	29.21	2,513.29	2,484.1	0.1056	8.9735
0.0020	17.5403	0.0010014	67.008	73.58	2,532.71	2,459.1	0.2611	8.7220
0.0030	24.1142	0.0010028	45.666	101.07	2,544.68	2,443.6	0.3546	8.5758
0.0040	28.9533	0.0010041	34.796	121.30	2,553.45	2,432.2	0.4221	8.4725
0.0050	32.8793	0.0010053	28.101	137.72	2,560.55	2,422.8	0.4761	8.3930
0.0060	36.1663	0.0010065	23.738	151.47	2,566.48	2,415.0	0.5208	8.3283
0.0070	38.9967	0.0010075	20.528	163.31	2,571.56	2,408.3	0.5589	8.2737
0.0080	41.5075	0.0010085	18.102	173.81	2,576.06	2,402.3	0.5924	8.2266
0.0090	43.7901	0.0010094	16.204	183.36	2,580.15	2,396.8	0.6226	8.1854
0.010	45.7988	0.0010103	14.673	191.76	2,583.72	2,392.0	0.6490	8.1481
0.015	53.9705	0.0010140	10.022	225.93	2,598.21	2,372.3	0.7548	8.0065
0.020	60.0650	0.0010172	7.6497	251.43	2,608.90	2,357.5	0.8320	7.9068
0.025	64.9726	0.0010198	6.2047	271.96	2,617.43	2,345.5	0.8932	7.8298
0.030	69.1041	0.0010222	5.2296	289.26	2,624.56	2,335.3	0.9440	7.7671
0.040	75.8720	0.0010264	3.9939	317.61	2,636.10	2,318.5	1.0260	7.6688
0.050	81.3388	0.0010299	3.2409	340.55	2,645.31	2,304.8	1.0912	7.5928
0.060	85.9496	0.0010331	2.7324	359.91	2,652.97	2,293.1	1.1454	7.5310
0.070	89.9556	0.0010359	2.3654	376.75	2,659.55	2,282.8	1.1921	7.4789
0.080	93.5107	0.0010385	2.0876	391.71	2,665.33	2,273.6	1.2330	7.4339
0.090	96.7121	0.0010409	1.8698	405.20	2,670.48	2,265.3	1.2696	7.3943
0.10	99.634	0.0010432	1.6943	417.52	2,675.14	2,257.6	1.3028	7.3589
0.12	104.810	0.0010473	1.4287	439.37	2,683.26	2,243.9	1.3609	7.2978
0.14	109.318	0.0010510	1.2368	458.44	2,690.22	2,231.8	1.4110	7.2462
0.16	113.326	0.0010544	1.09159	475.42	2,696.29	2,220.9	1.4552	7.2016
0.18	116.941	0.0010576	0.97767	490.76	2,701.69	2,210.9	1.4946	7.1623
0.20	120.240	0.0010605	0.88585	504.78	2,706.53	2,201.7	1.5303	7.1272
0.25	127.444	0.0010672	0.71879	535.47	2,716.83	2,181.4	1.6075	7.0528
0.30	133.556	0.0010732	0.60587	561.58	2,725.26	2,163.7	1.6721	6.9921
0.35	138.891	0.0010786	0.52427	584.45	2,732.37	2,147.9	1.7278	6.9407
0.40	143.642	0.0010835	0.46246	604.87	2,738.49	2,133.6	1.7769	6.8961
0.50	151.867	0.0010925	0.37486	640.35	2,748.59	2,108.2	1.8610	6.8214

https://doi.org/10.1515/9783111329703-016

(continued)

Pressure	Temperature	Specific volume		Specific enthalpy		Latent heat of vaporization	Specific entropy	
p (MPa)	t (°C)	Liquid	Vapor	Liquid	Vapor	r (kJ/kg)	Liquid	Vapor
		m³/kg	m³/kg	kJ/kg	kJ/kg		kJ/ (kg · K)	kJ/ (kg · K)
0.60	158.863	0.0011006	0.31563	670.67	2,756.66	2,086.0	1.9315	6.7600
0.70	164.983	0.0011079	0.27281	697.32	2,763.29	2,066.0	1.9925	6.7079
0.80	170.444	0.0011148	0.24037	721.20	2,768.86	2,047.7	2.0464	6.6625
0.90	175.389	0.0011212	0.21491	742.90	2,773.59	2,030.7	2.0948	6.6222
1.00	179.916	0.0011272	0.19438	762.84	2,777.67	2,014.8	2.1388	6.5859
1.10	184.100	0.0011330	0.17747	781.35	2,781.21	999.9	2.1792	6.5529
1.20	187.995	0.0011385	0.16328	798.64	2,784.29	985.7	2.2166	6.5225
1.30	191.644	0.0011438	0.15120	814.89	2,786.99	972.1	2.2515	6.4944
1.40	195.078	0.0011489	0.14079	830.24	2,789.37	959.1	2.2841	6.4683
1.50	198.327	0.0011538	0.13172	844.82	2,791.46	946.6	2.3149	6.4437
1.60	201.410	0.0011586	0.12375	858.69	2,793.29	934.6	2.3440	6.4206
1.70	204.346	0.0011633	0.11668	871.96	2,794.91	923.0	2.3716	6.3988
1.80	207.151	0.0011679	0.11037	884.67	2,796.33	911.7	2.3979	6.3781
1.90	209.838	0.0011723	0.104707	896.88	2,797.58	900.7	2.4230	6.3583
2.00	212.417	0.0011767	0.099588	908.64	2,798.66	890.0	2.4471	6.3395
2.20	217.289	0.0011851	0.090700	930.97	2,800.41	1869.4	2.4924	6.3041
2.40	221.829	0.0011933	0.083244	951.91	2,801.67	1,849.8	2.5344	6.2714
2.60	226.085	0.0012013	0.076898	971.67	2,802.51	1,830.8	2.5736	6.2409
2.80	230.096	0.0012090	0.071427	990.41	2,803.01	1,812.6	2.6105	6.2123
3.00	233.893	0.0012166	0.066662	1,008.2	2,803.19	1,794.9	2.6454	6.1854
3.50	242.597	0.0012348	0.057054	1,049.6	2,802.51	1,752.9	2.7250	6.1238
4.00	250.394	0.0012524	0.049771	1,087.2	2,800.53	1,713.4	2.7962	6.0688
5.00	263.980	0.0012862	0.039439	1,154.2	2,793.64	1,639.5	2.9201	5.9724
6.00	275.625	0.0013190	0.032440	1,213.3	2,783.82	1,570.5	3.0266	5.8885
7.00	285.869	0.0013515	0.027371	1,266.9	2,771.72	1,504.8	3.1210	5.8129
8.00	295.048	0.0013843	0.023520	1,316.5	2,757.70	1,441.2	3.2066	5.7430
9.00	303.385	0.0014177	0.020485	1,363.1	2,741.92	1,378.9	3.2854	5.6771
10.0	311.037	0.0014522	0.018026	1,407.2	2,724.46	1,317.2	3.3591	5.6139
11.0	318.118	0.0014881	0.015987	1,449.6	2705.34	1,255.7	3.4287	5.5525
12.0	324.715	0.0015260	0.014263	1,490.7	2,684.50	1,193.8	3.4952	5.4920
13.0	330.894	0.0015662	0.012780	1,530.8	2,661.80	1,131.0	3.5594	5.4318
14.0	336.707	0.0016097	0.011486	1,570.4	2,637.07	1,066.7	3.6220	5.3711
15.0	342.196	0.0016571	0.010340	1,609.8	2,610.01	1,000.2	3.6836	5.3091
16.0	347.396	0.0017099	0.009311	1,649.4	2,580.21	930.8	3.7451	5.2450
17.0	352.334	0.0017701	0.008373	1,690.0	2,547.01	857.1	3.8073	5.1776
18.0	357.034	0.0018402	0.007503	1,732.0	2,509.45	777.4	3.8715	5.1051
19.0	361.514	0.0019258	0.006679	1,776.9	2,465.87	688.9	3.9395	5.0250
20.0	365.789	0.0020379	0.005870	1,827.2	2,413.05	585.9	4.0153	4.9322

(continued)

Pressure	Temperature	Specific volume		Specific enthalpy		Latent heat of vaporization	Specific entropy	
p (MPa)	t (°C)	Liquid	Vapor	Liquid	Vapor	r (kJ/kg)	Liquid	Vapor
		m³/kg	m³/kg	kJ/kg	kJ/kg		kJ/ (kg · K)	kJ/ (kg · K)
21.0	369.868	0.0022073	0.005012	1,889.2	2,341.67	452.4	4.1088	4.8124
22.0	373.752	0.0027040	0.003684	2013.0	2,084.02	71.0	4.2969	4.4066
22.064	373.99	0.003106	0.003106	2085.9	2,085.9	0.0	4.4092	4.4092

Source of data: Tables A1 and A2 are generated using the Engineering Equation Solver (EES) software developed by S. A. Klein and F. L. Alvarado. The routine used in calculations is the highly accurate Steam_IAPWS, which incorporates the 1995 Formulation for the Thermodynamic Properties of Ordinary Water Substance for General and Scientific Use, issued by the International Association for the Properties of Water and Steam (IAPWS). This formulation replaces the 1984 formulation of Haar, Gallagher, and Kell (NBS/NRC Steam Tables, Hemisphere Publishing Co., 1984), which is also available in EES as the routine STEAM. The new formulation is based on the correlations of Saul and Wagner (J. Phys. Chem. Ref. Data, 16, 893, 1987) with modifications to adjust to the International Temperature Scale of 1990. The modifications are described by Wagner and Pruss (J. Phys. Chem. Ref. Data, 22, 783, 1993).

Appendix
Table A3: Unsaturated water and superheated water vapor

Pressure	0.001 MPa(t_s = 6.949 °C)			0.005 MPa(t_s = 32.879 °C)		
p(MPa)	v'	h'	s'	v'	h'	s'
	0.001001 m³/kg	29.21 kJ/kg	0.1056 kJ/(kg·K)	0.0010053 m³/kg	137.72 kJ/kg	0.4761 kJ/(kg·K)
	v''	h''	s''	v''	h''	s''
	129.185 m³/kg	2,513.3 kJ/kg	8.9735 kJ/(kg·K)	28.191 m³/kg	2,560.6 kJ/kg	8.3930 kJ/(kg·K)
t °C	v m³/kg	h kJ/kg	s kJ/(kg·K)	v m³/kg	h kJ/kg	s kJ/(kg·K)
0	0.001002	−0.05	−0.0002	0.0010002	−0.05	−0.0002
10	130.598	2,519.0	8.9938	0.0010003	42.01	0.1510
20	135.226	2,537.7	9.0588	0.0010018	83.87	0.2963
40	144.475	2,575.2	9.1823	28.854	2,574.0	8.4366
60	153.717	2,612.7	9.2984	30.712	2,611.8	8.5537
80	162.956	2,650.3	9.4080	32.566	2,649.7	8.6639
100	172.192	2,688.0	9.5120	34.418	2,687.5	8.7682
120	181.426	2,725.9	9.6109	36.269	2,725.5	8.8674
140	190.660	2,764.0	9.7054	38.118	2,763.7	8.9620
160	199.893	2,802.3	9.7959	39.967	2,802.0	9.0526
180	209.126	2,840.7	9.8827	41.815	2,840.5	9.1396
200	218.358	2,879.4	9.9662	43.662	2,879.2	9.2232
220	227.590	2,918.3	10.0468	45.510	2,918.2	9.3038
240	236.821	2,957.5	10.1246	47.357	2,957.3	9.3816
260	0.001002	−0.05	−0.0002	0.0010002	−0.05	−0.0002
280	130.598	2,519.0	8.9938	0.0010003	42.01	0.1510
300	135.226	2,537.7	9.0588	0.0010018	83.87	0.2963
350	144.475	2,575.2	9.1823	28.854	2,574.0	8.4366
400	153.717	2,612.7	9.2984	30.712	2,611.8	8.5537
450	162.956	2,650.3	9.4080	32.566	2,649.7	8.6639
500	172.192	2,688.0	9.5120	34.418	2,687.5	8.7682
550	181.426	2,725.9	9.6109	36.269	2,725.5	8.8674
600	190.660	2,764.0	9.7054	38.118	2,763.7	8.9620

https://doi.org/10.1515/9783111329703-017

Pressure	0.010 MPa(t_s = 45.799 °C)			0.10 MPa(t_s = 99.634 °C)		
p(MPa)	v'	h'	s'	v'	h'	s'
	0.0010103 m³/kg	191.76 kJ/kg	0.6490 kJ/(kg·K)	0.0010431 m³/kg	417.52 kJ/kg	1.3028 kJ/(kg·K)
	v''	h''	s''	v''	h''	s''
	14.673 m³/kg	2,583.7 kJ/kg	8.1481 kJ/(kg·K)	1.6943 m³/kg	2,675.1 kJ/kg	7.3589 kJ/(kg·K)
t °C	v m³/kg	h kJ/kg	s kJ/(kg·K)	v m³/kg	h kJ/kg	s kJ/(kg·K)
0	0.0010002	−0.04	−0.0002	0.0010002	0.05	−0.0002
10	0.0010003	42.01	0.1510	0.0010003	42.10	0.1510
20	0.0010018	83.87	0.2963	0.0010018	83.96	0.2963
40	0.0010079	167.51	0.5723	0.0010078	167.59	0.5723
60	15.336	2,610.8	8.2313	0.0010171	251.22	0.8312
80	16.268	2,648.9	8.3422	0.0010290	334.97	1.0753
100	17.196	2,686.9	8.4471	1.6961	2,675.9	7.3609
120	18.124	2,725.1	8.5466	1.7931	2,716.3	7.4665
140	19.050	2,763.3	8.6414	1.8889	2,756.2	7.5654
160	19.976	2,801.7	8.7322	1.9838	2,795.8	7.6590
180	20.901	2,840.2	8.8192	2.0783	2,835.3	7.7482
200	21.826	2,879.0	8.9029	2.1723	2,874.8	7.8334
220	22.750	2,918.0	8.9835	2.2659	2,914.3	7.9152
240	23.674	2,957.1	9.0614	2.3594	2,953.9	7.9940
260	24.598	2,996.5	9.1367	2.4527	2,993.7	8.0701
280	25.522	3,036.2	9.2097	2.5458	3,033.6	8.1436
300	26.446	3,076.0	9.2805	2.6388	3,073.8	8.2148
350	28.755	3,176.6	9.4488	2.8709	3174.9	8.3840
400	31.063	3,278.7	9.6064	3.1027	3,277.3	8.5422
450	33.372	3,382.3	9.7548	3.3342	3,381.2	8.6909
500	35.680	3,487.4	9.8953	3.5656	3,486.5	8.8317
550	37.988	3,594.3	10.0293	3.7968	3,593.5	8.9659
600	40.296	3,703.4	10.1579	4.0279	3,702.7	9.0946

Pressure	0.5MPa(t_s = 151.867 °C)			1MPa(t_s = 179.916 °C)		
p(MPa)	v′	h′	s′	v′	h′	s′
	0.0010925	640.35	1.8610	0.0011272	762.84	2.1388
	m³/kg	kJ/kg	kJ/(kg·K)	m³/kg	kJ/kg	kJ/(kg·K)
	v″	h″	s″	v″	h″	s″
	0.37490	2,748.60	6.8214	0.19440	2,777.7	6.5859
	m³/kg	kJ/kg	kJ/(kg·K)	m³/kg	kJ/kg	kJ/(kg·K)
t	v	h	s	v	h	s
°C	m³/kg	kJ/kg	kJ/(kg·K)	m³/kg	kJ/kg	kJ/(kg·K)
0	0.0010000	0.46	−0.0002	0.0009997	0.97	0.0001
10	0.0010001	42.49	0.1510	0.0009999	42.98	0.1509
20	0.0010016	84.33	0.2963	0.00180014	84.80	0.2961
40	0.0010077	167.94	0.5721	0.0010074	168.38	0.5719
60	0.0010169	251.56	0.8310	0.0010167	251.98	0.8307
80	0.0010288	335.29	1.0750	0.0010286	335.69	1.0747
100	0.0010432	419.36	1.3066	0.0010430	419.74	1.3062
120	0.0010601	503.97	1.5275	0.0010599	504.32	1.5270
140	0.0010796	589.30	1.7392	0.0010783	589.62	1.7386
160	0.38358	2,767.2	6.8647	0.0011017	675.84	1.9424
180	0.40450	2,811.7	6.9651	0.19443	2,777.9	6.5864
200	0.42487	2,854.9	7.0585	0.20590	2,827.3	6.6931
220	0.44485	2,897.3	7.1462	0.21686	2,874.2	6.7903
240	0.46455	2,939.2	7.229.5	0.22745	2,919.6	6.880.4
260	0.484.04	2,980.8	7.3091	0.23779	2,963.8	6.9650
280	0.50336	3,022.2	7.3853	0.24793	3,007.3	7.0451
300	0.52255	3063.6	7.4588	0.25793	3,050.4	7.1216
350	0.57012	3,167.0	7.6319	0.28247	3,157.0	7.2999
400	0.61729	3,271.1	7.7924	0.30658	3,263.1	7.4638
420	0.63608	3,312.9	7.8537	0.31615	3,305.6	7.5260
440	0.65483	3,354.9	7.9135	0.32568	3,348.2	7.5866
450	0.66420	3,376.0	7.9428	0.33043	3,369.6	7.6163
460	0.67356	3,3972	7.97119	0,33518	3,390.9	7.6456
480	0.69226	3,439.6	8.0289	0.34465	3,433.8	7.7033
500	0.71094	3,482.2	8.0848	0.35410	3,476.8	7.7597
550	0.75755	3,589.9	8.2198	0.37764	3,585.4	7.8958
600	0.80408	3,699.6	8.3491	0.40109	3,695.71	8.0259

Pressure	3 MPa(t_s = 233.893 °C)			5 MPa(t_s = 263.980 °C)		
p (MPa)	v'	h'	s'	v'	h'	s'
	0.001216 m³/kg	1008.2 kJ/kg	2.6454 kJ/(kg·K)	0.0012861 m³/kg	1,154.2 kJ/kg	2.9200 kJ/(kg·K)
	v''	h''	s''	v''	h''	s''
	0.066700 m³/kg	2,803.2 kJ/kg	6.1854 kJ/(kg·K)	0.039400 m³/kg	2,793.6 kJ/kg	5.9724 kJ/(kg·K)
t °C	v m³/kg	h kJ/kg	s kJ/(kg·K)	v m³/kg	h kJ/kg	s kJ/(kg·K)
0	0.0009887	3.01	0.0000	0.0009977	5.04	0.0002
10	0.0009989	44.92	0.1507	0.0009979	46.87	0.1506
20	0.0010005	86.68	0.2957	0.0009996	88.55	0.2952
40	0.0010066	170.15	0.5711	0.0010057	171.92	0.5704
60	0.0010158	253.66	0.8296	0.0010149	255.34	0.8266
80	0.0010276	377.28	1.0734	0.0010267	338.87	1.0721
100	0.0010420	421.24	1.3047	0.0010410	422.75	1.3031
120	0.0010587	505.73	1.5252	0.0010576	507.14	1.5234
140	0.0010781	590.92	1.7366	0.0010768	592.23	1.7345
160	0.0011002	677.01	1.9400	0.0010988	678.19	1.9377
180	0.0011256	764.23	2.1369	0.0011240	765.25	2.1342
200	0.0011549	852.93	2.3284	0.0011529	853.75	2.3253
220	0.0011891	943.65	2.5162	0.0011867	944.21	2.5125
240	0.068184	2,823.4	6.2250	0.0012266	1,037.3	2.6976
260	0.072828	2,884.4	6.3417	0.0012751	1,134.3	2.8829
280	0.077101	2,940.1	6.4443	0.042228	2,855.8	6.0864
300	0.084191	2,992.4	6.5371	0.045301	2,923.3	6.2064
350	0.090520	3,114.4	6.7414	0.051932	3,067.4	6.4477
400	0.099352	3,230.1	6.9199	0.057804	3,194.9	6.6446
420	0.102787	3,275.4	6.9864	0.060033	3,243.6	6.7159
440	0.106180	3,320.5	7.0505	0.062216	3,291.5	6.7840
450	0.107864	3,343.0	7.0817	0.063291	3,315.2	6.8170
460	0.109540	3,365.4	7.1125	0.064358	3,338.8	6.8494
480	0.112870	3,410.1	7.1728	0.066469	3,385.6	6.9125
500	0.116174	3,454.9	7.2314	0.068552	3,432.2	6.9735
550	0.124349	3,566.9	7.3718	0.073664	3,548.0	7.1187
600	0.132427	3,679.9	7.5051	0.078675	3,663.9	7.2553

Pressure	7 MPa(t_s = 285.869 °C)			10 MPa(t_s = 311.037 °C)		
p (MPa)	v'	h'	s'	v'	h'	s'
	0.0013515 m³/kg	1,266.9 kJ/kg	3.1210 kJ/(kg·K)	0.0014522 m³/kg	1,407.2 kJ/kg	3.3591 kJ/(kg·K)
	v''	h''	s''	v''	h''	s''
	0.027400 m³/kg	2,771.7 kJ/kg	5.8129 kJ/(kg·K)	0.018000 m³/kg	2,724.5 kJ/kg	5.6139 kJ/(kg·K)
t °C	v m³/kg	h kJ/kg	s kJ/(kg·K)	v m³/kg	h kJ/kg	s kJ/(kg·K)
0	0.0009967	7.07	0.0003	0.0009952	10.09	0.0004
10	0.0009970	48.80	0.1504	0.0009956	51.70	0.1500
20	0.0009986	90.42	0.2948	0.0009973	93.22	0.2942
40	0.0010048	173.69	0.5696	0.0010035	176.34	0.5684
60	0.0010140	257.01	0.8275	0.0010127	259.53	0.8259
80	0.0010258	340.46	1.0708	0.0010244	342.85	1.0688
100	0.0010399	424.250	1.3016	0.0010385	426.51	1.2993
120	0.0010565	508.55	1.5216	0.0010549	510.68	1.5190
140	0.0010756	593.540	1.7325	0.0010738	595.50	1.7294
160	0.0010974	679.370	1.9353	0.0010953	681.16	1.9319
180	0.0011223	766.280	2.1315	0.0011199	767.84	2.1275
200	0.0011510	854.59	2.3222	0.0011481	855.88	2.3176
220	0.0011842	944.79	2.5089	0.0011807	945.71	2.5036
240	0.0012235	1,037.6	2.6933	0.0012190	1,038.0	2.6870
260	0.0012710	1,134.0	2.8776	0.0012650	1,133.6	2.8698
280	0.0013307	1,235.7	3.0648	0.0013222	1,234.2	3.0549
300	0.029457	2,837.5	5.9291	0.0013975	1,342.3	3.2469
350	0.035225	3,014.8	6.2265	0.022415	2,922.1	5.9423
400	0.039917	3,157.3	6.4465	0.026402	3,095.8	6.2109
450	0.044143	3,286.2	6.6314	0.029735	3,240.5	6.4184
500	0.048110	3,408.9	6.7954	0.032750	3,372.8	6.5954
520	0.049649	3,457.0	6.8569	0.033900	3,423.8	6.6605
540	0.051166	3,504.8	6.9164	0.035027	3,474.1	6.7232
550	0.051917	3,528.7	6.9456	0.035582	3,499.1	6.7537
560	0.052664	3,552.4	6.9743	0.036133	3,523.9	6.7837
580	0.054147	3,600.0	7.0306	0.037222	3,573.3	6.8423
600	0.055617	3,647.5	7.0857	0.038297	3,622.5	6.8992

Pressure	14.0 MPa(t_a = 336.707 °C)			20.0 MPa(t_s = 365.789 °C)		
p (MPa)	v′	h′	s′	v′	h′	s′
	0.0016097 m³/kg	1,570.4 kJ/kg	3.6220 kJ/(kg·K)	0.0020379 m³/kg	1,827.2 kJ/kg	4.0153 kJ/(kg·K)
	v″	h″	s″	v″	h″	s″
	0.011500 m³/kg	2,637.1 kJ/kg	5.3711 kJ/(kg·K)	0.0058702 m³/kg	2,413.1 kJ/kg	4.9322 kJ/(kg·K)
t °C	v m³/kg	h kJ/kg	s kJ/(kg·K)	v m³/kg	h kJ/kg	s kJ/(kg·K)
0	0.0009933	14.10	0.0005	0.0009904	20.08	0.0006
10	0.0009938	55.55	0.1496	0.0009911	61.29	0.1488
20	0.0009955	96.95	0.2932	0.0009929	102.50	0.2919
40	0.0010018	179.86	0.5669	0.0009992	185.13	0.5645
60	0.0010109	262.88	0.8239	0.0010084	267.90	0.8207
80	0.0010226	346.04	1.0663	0.0010199	350.82	1.0624
100	0.0010365	429.53	1.2962	0.0010336	434.06	1.2917
120	0.0010527	513.52	1.5155	0.0010496	517.79	1.5103
140	0.0010714	598.14	1.7254	0.0010679	602.12	1.7195
160	0.0010926	683.56	1.9273	0.0010886	687.20	1.9206
180	0.0011167	769.96	2.1223	0.0011121	773.19	2.1147
200	0.0011443	857.63	2.3116	0.0011389	860.36	2.3029
220	0.0011761	947.00	2.4966	0.0011695	949.07	2.4865
240	0.0012132	1,038.6	2.6788	0.0012051	1,039.8	2.6670
260	0.0012574	1,133.4	2.8599	0.0012469	1,133.4	2.8457
280	0.0013117	1,232.5	3.0424	0.0012974	1,230.7	3.0249
300	0.013814	1,338.2	3,2300	0.0013605	1,333.4	3.2072
350	0.013218	2,751.2	5.5564	0.0016645	1,645.3	3.7275
400	0.017218	3,001.1	5.9436	0.0099458	2,816.8	5.5520
450	0.020074	3,174.2	6.1919	0.0127013	3,060.7	5.9025
500	0.022512	3,322.3	6.3900	0.0147681	3,239.3	6.1415
520	0.023418	3,377.9	6.4610	0.0155046	3,303.0	6.2229
540	0.024295	3,432.1	6.5285	0.0162067	3,364.0	6.2989
550	0.024724	3,458.7	6.5611	0.0165471	3,393.7	6.3352
560	0.025147	3,485.2	6.5931	0.0168811	3,422.9	6.3705
580	0.025978	3,537.5	6.6551	0.0175328	3,480.3	6.4385
600	0.026792	3,589.1	6.7149	0.0181655	3,536.3	6.5035

Pressure	25 MPa			30 MPa		
t °C	v m³/kg	h kJ/kg	s kJ/(kg·K)	v m³/kg	h kJ/kg	s kJ/(kg·K)
0	0.0009880	25.01	0.0006	0.0009857	29.92	0.0005
10	0.0009888	66.04	0.1481	0.0009866	70.77	0.1474
20	0.0009908	107.11	0.2907	0.0009887	111.71	0.2895
40	0.0009972	189.51	0.5626	0.0009951	193.87	0.5606
60	0.0010063	272.08	0.8182	0.0010042	276.25	0.8156
80	0.0010177	354.80	1.0593	0.0010155	358.78	1.0562
100	0.0010313	437.85	1.2880	0.0010290	441.64	1.2844
120	0.0010470	521.36	1.5061	0.0010445	524.95	1.5019
140	0.0010650	605.46	1.7147	0.0010622	608.82	1.7100
160	0.0010854	690.27	1.9152	0.0010822	693.36	1.9098
180	0.0011084	775.94	2.1085	0.0011048	778.72	2.1024
200	0.0011345	862.71	2.2959	0.0011303	865.12	2.2890
220	0.0011643	950.91	2.4785	0.0011593	952.85	2.4706
240	0.0011986	1,041.0	2.6575	0.0011925	1,042.3	2.6485
260	0.0012387	1,133.6	2.8346	0.0012311	1,134.1	2.8239
280	0.0012866	1,229.6	3.0113	0.0012766	1,229.0	2.9985
300	0.0013453	1,330.3	3.1901	0.0013317	1,327.9	3.1742
350	0.0015981	1,623.0	3.6788	0.0015522	1,608.0	3.6420
400	0.0060014	2,578.0	5.1386	0.0027929	2,150.6	4.4721
450	0.0091666	2,950.5	5.6754	0,0067363	2,822.1	5.4433
500	0.0111229	3,164.1	5.9614	0.0086761	3,083.3	5.7934
520	0.0117897	3,236.1	6.0534	0.0093033	3,165.4	5.8982
540	0.0124156	3,303.8	6.1377	0.0098825	3,240.8	5.9921
550	0.0127161	3,336.4	6,1775	0.0101580	3,276.6	6.0359
560	0.0130095	3,368.2	6.2160	0.0104254	3,311.4	6.0780
580	0.0135778	3,430.2	6.2895	0.0109397	3,378.5	6.157 6
600	0.0141249	3,490.2	6.3591	0.0114310	3,442.9	6.2321

Description: Above the thick horizontal line is unsaturated water, below the thick horizontal line is superheated water vapor.

References

[1]   BP Statistical Yearbook of World Energy, 2021.
[2]   L. X. Fan, Z. K. Tu, and S. H. Chan. Recent Development of Hydrogen and Fuel Cell Technologies: A Review. Energy Reports, 2021, 7:8421–8446.
[3]   C. Rayment and S. Sherwin. Introduction to Fuel Cell Technology. 2003.
[4]   S. Mekhilef, R. Saidur, A. Safari. Comparative study of different fuel cell technologies. Renewable and Sustainable Energy Reviews. 2012, 16(1):981–989.
[5]   B. Cook. An introduction to fuel cells and hydrogen technology. Engineering Science and Education Journal, 2003, 11(6):205–216.
[6]   K. Xie. Reviews of clean coal conversion technology in china: situations & challenges. Chinese Journal of Chemical Engineering. 2021, 35(7):62–69.
[7]   N. Amin. Solar photovoltaic technologies: from inception toward the most reliable energy resource. Encyclopedia of Sustainable Technologies, 2017, 11–26.
[8]   W. F. E. Feller. Air Compressors: Their Installation, Operation, and Maintenance. New York: McGraw-Hill, 1944.
[9]   A. Bejan. Entropy Generation Through Heat and Fluid Flow. New York: Wiley Interscience, 1982.
[10] H. Sorensen. Energy Conversion Systems. New York: John Wiley & Sons, 1983.
[11] A. Bejan. Advanced Engineering Thermodynamics. 3rd ed. New York: Wiley Interscience, 2006.
[12] M. Kostic. Analysis of Enthalpy Approximation for Compressed Liquid Water. ASME Journal of Heat Transfer, 2006, 128:421–426.
[13] C. R. Ferguson and A. T. Kirkpatrick. Internal Combustion Engines: Applied Thermosciences, 2nd ed. New York: Wiley, 2000.
[14] W. Pulkrabek. Engineering Fundamentals of the Internal Combustion Engine, 2nd ed. Upper Saddle River, NJ: Prentice-Hall, 2004.
[15] ASHRAE Handbook of Refrigeration, SI version. Atlanta, GA: American Society of Heating, Refrigerating, and Air-Conditioning Engineers, Inc, 1994.
[16] W. F. Stoecker and J. W. Jones. Refrigeration and Air Conditioning. 2nd ed. New York: McGraw-Hill, 1982.
[17] Heat Pump Systems – A Technology Review. OECD Report, Paris, 1982.
[18] A. Bejan. Advanced Engineering Thermodynamics. 3rd ed. New York: Wiley, 2006.
[19] K. Wark, Jr. Advanced Thermodynamics for Engineers. New York: McGraw-Hill, 1995.
[20] B. Chapman. Heat transfer. Magnetic Resonance Materials in Biology Physics & Medicine, 1974, 9(3):146–151. DOI: 10.1016/s1352-8661(99)00067-8.
[21] M. I. S. Pe. Heat Transfer Theory. Surface Production Operations. 3rd ed, 2014, 39–97.
[22] Y. A. Çengel, M. A. Boles, and M. Kanoğlu. Thermodynamics: An Engineering Approach. McGraw-Hill, 2019.
[23] M. W. Zemansky. Heat and Thermodynamics: An Intermediate Textbook for Students of Physics, Chemistry, and Engine. McGraw-Hill, 1997.

https://doi.org/10.1515/9783111329703-018

Index

https://doi.org/10.1515/9783111329703-019

Also of interest

Thermal Analysis and Thermodynamics
In Materials Science
Detlef Klimm, 2022
ISBN 978-3-11-074377-7, e-ISBN 978-3-11-074378-4, ePUB 978-3-11-074384-5

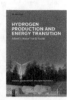

Volume 1 Hydrogen Production and Energy Transition
Marcel Van de Voorde (Ed.), 2021
ISBN 978-3-11-059622-9, e-ISBN 978-3-11-059625-0, ePUB 978-3-11-059405-8

Mass, Momentum and Energy Transport Phenomena
A Consistent Balances Approach
2nd Edition
Harry Van den Akker and Robert F. Mudde, 2023
ISBN 978-3-11-124623-9, e-ISBN 978-3-11-124657-4, ePUB 978-3-11-124715-1

Multiphase Reactors
Reaction Engineering Concepts, Selection, and Industrial Applications
Jan Harmsen and René Bos, 2023
ISBN 978-3-11-071376-3, e-ISBN 978-3-11-071377-0, ePUB 978-3-11-071384-8

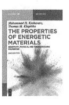

The Properties of Energetic Materials
Sensitivity, Physical and Thermodynamic Properties
2nd Edition
Mohammad Hossein Keshavarz and Thomas M. Klapötke, 2021
ISBN 978-3-11-074012-7, e-ISBN 978-3-11-074015-8, ePUB 978-3-11-074024-0

www.degruyter.com

Printed in the USA
CPSIA information can be obtained
at www.ICGtesting.com
LVHW082036140524
780250LV00005B/733

9 783111 329697